LIFE AT THE EDGE OF SIGHT

LIFE AT THE EDGE OF SIGHT

A Photographic Exploration of the Microbial World

SCOTT CHIMILESKI

ROBERTO KOLTER

THE BELKNAP PRESS *of* HARVARD UNIVERSITY PRESS

Cambridge, Massachusetts & London, England · 2017

First printing

Library of Congress Cataloging-in-Publication Data

Names: Chimileski, Scott, author. | Kolter, Roberto, 1953– author.
Title: Life at the edge of sight : a photographic exploration of the microbial
    world / Scott Chimileski and Roberto Kolter.
Description: Cambridge, Massachusetts : The Belknap Press of Harvard
    University Press, 2017. | Includes bibliographical references and index.
Identifiers: LCCN 2017019203 | ISBN 9780674975910 (alk. paper)
Subjects: LCSH: Microorganisms. | Microorganisms—Pictorial works. | LCGFT: Photobooks.
Classification: LCC QR54 .C45 2017 | DDC 579.022/2—dc23 LC record available at
    https://lccn.loc.gov/2017019203

CONTENTS

FOREWORD

Elio Schaechter

CONTEMPLATING THE MICROBIAL WORLD requires us to reboot our brains. How else can we deal with its numerous tiny members, numbering well over 10 followed by 30 zeroes? Or with the bewildering variety of their genomes—well into the millions? When considering what microbes do, it's easier to ask what they don't do. They have transformed this planet—its geology, its atmosphere, and its climate. They are essential to life and to its evolution. This is indubitably the planet of the microbes, and we would do well to recognize it. If we think of microbes at all, we usually think of them as "germs." Germs do indeed cause disease, but assigning to pathogenic microbes a major role in human affairs is as anthropocentric as believing that Earth is the center of the universe.

No single book can do justice to the vastness of microbial experience. But it can act as an ambassador, sharing stories that illuminate that otherwise unseen world. This is what readers will find here. The authors' grand tour introduces readers to microbes with engaging tales, each introducing a foundational concept or two.

The authors come to the task with different experiences: one is a well-known researcher and teacher whose highly significant contributions have spanned the world of microbes, the other a younger member of the profession with a true passion and skill for sharing his field with the general public. Together, they have found a delightful voice with which to present their approach to this sometimes astonishing and always captivating world.

LIFE ON EARTH HAD ITS ORIGINS about four billion years ago. The early life-forms were most certainly microbes—far too small to see with the naked eye. For three billion years, microbes reigned supreme on the planet. During this time they evolved to adapt to the cooling Earth's emerging environments and began their colonization of virtually every corner of what today we call the biosphere. Microbes made that biosphere. In doing so, they not only achieved unimaginable levels of species diversity but shaped the very Earth they were colonizing. They made and broke rocks, gave rise to the oxygen in the atmosphere, and participated in many other geological processes. Large organisms—the plants and animals—did not begin to populate the planet until about 500 million years ago, during the Cambrian explosion. Our own human evolution took place in environments where we were constantly surrounded by microbes. Not only did microbes shape our evolution, we in turn shaped their evolution. From the outset, animal life formed intimate symbioses with the microbes surrounding them. The microbes evolved to adapt to the new environments afforded by these newly emerging large organisms.

Part of what makes us human is our long record of domesticating species to our benefit. We tend to think of the domestication of plants and animals as a key development in the establishment of civilization. The domestication of microbes to produce diverse foods and beverages—which we consume to this day—likely began even earlier and has been equally important in our history. For the most part, however, microbes escaped our notice because they are so small. Even though the interaction of humans with microbes has driven our evolution, it was not until relatively recently that we recognized the existence of microbes as living entities.

The discipline of microbiology has a relatively short history. It was not until the seventeenth century that the Dutchman Antoni van Leeuwenhoek first opened the doors to this previously unknown universe of microscopic life. By the end of the nineteenth century, the foundations of the science had been laid by three pioneering microbiologists: Louis Pasteur, Martinus Beijerinck, and Sergei Winogradsky. Pasteur developed sterile techniques and determined that the alcoholic fermentation involved in wine production resulted from the activity of a living microbe—brewer's yeast. Beijerinck and Winogradsky discovered that microbes played a role in the global cycling of nitrogen from inert atmospheric gas to its nutritious forms, ammonia and nitrate. What characterized the work of these three pioneering microbiologists was their realization that microbial activities both influenced and were influenced by the microbes' local environment. In that sense these scientists were microbial ecologists. Although they all recognized that microbes could be causative agents of disease, their thinking was not dominated by that concept.

It is difficult to overstate the influence of Robert Koch on the science and medicine of the twentieth century. A pioneer microbiologist, Koch developed techniques to obtain pure cultures of microbes, cultures where one and only one microbial species was present. These pure cultures allowed him to prove that an individual species of microbe caused a specific infectious disease. By isolating the microbe from diseased animals, growing it as a pure culture, reinfecting a healthy animal, seeing the development of the same disease, and then reisolating the same microbe from this second diseased animal, Koch set the gold standard for defining a causative agent of any given infectious disease. This standard has stood the test of time, and it gave us a whole new way to see microbes.

Throughout history, plagues and pandemics have decimated both human and animal populations, yet the causes of these diseases remained a mystery. Perhaps the most devastating of those recorded in Europe was the Black Death, which struck in the middle of the fourteenth century, killing about half of the population. There is no doubt that the Black Death and subsequent regional outbreaks of this disease over the following centuries influenced the course of humanity deeply.

Many other pandemics have affected and continue to affect us. But by the early twentieth century, the causes of these infectious diseases were beginning to be identified, largely thanks to the work of Robert Koch. By the time the world population was reduced by the 1918 influenza virus pandemic, there was a clear sense that a microbe was involved in causing the devastation. What followed was a period of intense research, aimed not only at finding the microbe responsible but also at finding cures for other infectious diseases. By the middle of the century, these investigations resulted in the development of antibiotics as therapeutic agents to treat bacterial infections.

However, the effect that epidemics have on the human psyche and the great advances made in treating infections with antibiotics had some unintended negative consequences with regard to our perceptions of microbes. The idea of bacteria and microbes became associated with disease, and the general sense was that, as such, microbes should be eradicated. Thus, for a good part of the twentieth century microbiology was dominated by the concept that microbes cause disease and that they do so under all conditions. The positive aspects of microbes and the appreciation of the roles that they play in maintaining healthy ecosystems became rather obscured.

A better understanding of microbes was reached during the last decades of the twentieth century. Numerous findings led to the recognition that the microbial world is composed of a remarkably large number of different species, and yet only a few dozen are known to cause human disease. In addition, the ability of those human pathogens to cause disease is often controlled by which other microbes are present—by the local ecology. The emerging view is that microbes play very important and mostly beneficial roles in almost every aspect of life on the planet, with the added recognition that science has just barely scratched the surface in studying the mysteries of the microbial world.

Enough is known today to appreciate much of the beauty that is inherent in visualizing the life of microbes. We have been so taken by this beauty and by the beneficial aspects of microbial life that we felt compelled to assemble this book that melds stories and images. The sheer number and limitless variety of microbes means that our work could not be comprehensive. Rather, our aim is to share our excitement, to show readers a sample of microbial biodiversity, and to communicate and advance some fundamental aspects of microbial science.

The chapters alternate between stories that take the reader back in time to gaze over the shoulders of historic scientists and stories that follow fictional characters exploring the microbial world today. Certain themes resonate through all chapters. Some are quite concrete, including the ways microbes impact humans, the hidden forests of microbes all around us, and the collective behaviors of microbes. We will also contemplate the balance between art and science, and the balance of reductionism and synthesis as forces in the scientific process. Finally, we will take a trip down the rabbit hole to think about how we experience microscopic and macroscopic size scales, to compare and contrast microbial worlds and cosmic worlds, and to contemplate the interconnectivity of life on Earth.

LIFE AT THE EDGE OF SIGHT

1

From an Ancient Chalk Graveyard

IN 1668, ANTONI VAN LEEUWENHOEK stood on the coast of England, fascinated by a white chalk cliff. How could this young man from Holland have known that, centuries later, he would be remembered as one of the most celebrated scientists of all time? He could not have realized that a homemade magnifying lens tucked in his luggage was the most powerful microscope on Earth at the time. He was simply following his curiosity.

Antoni picked up a crumble of chalk from beside the cliff and set it on the pin of his metal microscope, holding it against his face and up to the Sun. As he carefully rotated a tiny focus screw, "very small transparent particles" appeared on the surface of the chalk, "lying one upon another." The instant that light came through the glass lens and energized the photoreceptors in his eye was one of the greatest moments in the history of science. He was not looking at living organisms. Rather, he was looking at particles from organisms that lived in the past, a trace of an entire microscopic world that he would soon discover. It was Antoni's first glimpse of life beyond the edge of sight. No Egyptian pharaoh, no Greek philosopher, not Leonardo da Vinci, not Sir Isaac Newton, not Galileo Galilei, no one had ever seen a biological entity of this size.

If we look at chalk today from the White Cliffs of Dover in England, we can locate the same transparent particles. We can also use modern microscopes to zoom in closer and see much more. Antoni's simple light microscopes created images that to his eyes were 70 to 300 times larger than the actual size of the objects. It was technology that bordered on magic in the seventeenth century.

A scanning electron microscope can magnify an object well over 100,000 times. By directing a beam of electrons against the chalk and analyzing how the electrons bounce off the surface to create an image, we see a mysterious landscape of disc-shaped objects. What are these structures? The discs are themselves

The White Cliffs of Dover on the coast of England.

1

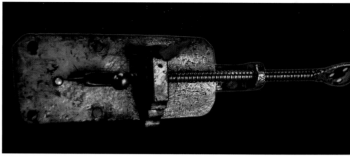

Crumbles of chalk from the White Cliffs.

A replica of Antoni van Leeuwenhoek's microscope. The real ones reside in the Museum Boerhaave, Leiden, the Netherlands.

(Opposite) Mysterious disc-shaped objects and smaller fragments are seen when chalk is analyzed using a scanning electron microscope.

(Overleaf) Light blue and green patches in the North Atlantic Ocean beneath the clouds. Newfoundland and Greenland are on the left and Europe is on the right.

formed by smaller bony fragments, intricately lined up one upon another. Back in 1668, Antoni could not see the discs in such magnificent detail. On the day that he first examined the chalk, the full story of this ancient graveyard remained locked in the cliffs.

Marvelous though they may be, the most magnified images of chalk particles attainable with a scanning electron microscope would not have provided enough information to satisfy Antoni's curiosity. What is chalk? To answer this question, we need more context. We need to leave the chalk cliffs on the coast of England entirely.

Imagine we are thousands of kilometers in the sky overlooking the North Atlantic Ocean. From this fantastic vantage point, let us ask ourselves: what are the most obvious signals of life that we can detect on Earth from afar? We see green where there are forests over land. Sooner or later we might spot a pod of great blue whales, the largest animals on this planet, past or present, migrating across the ocean. But long before this, our eyes are drawn to light blue patches beneath the white clouds. Moving closer to look through a break in the clouds we wonder: what are these brilliant patterns in the ocean that swirl across thousands of kilometers?

The patterns and colors are not caused by any great wild beast whose skeleton you might find suspended from the ceiling of a natural history museum. They are caused by photosynthetic plankton that wander across the seas. They are microbes—organisms that are too small to see with the naked eye. The visual acuity of the human eye allows us to see objects about the width of a human hair and larger. Each algal plankton cell is a minute sphere one-tenth the width of a human hair or smaller in diameter. How can this be? One of the most noticeable

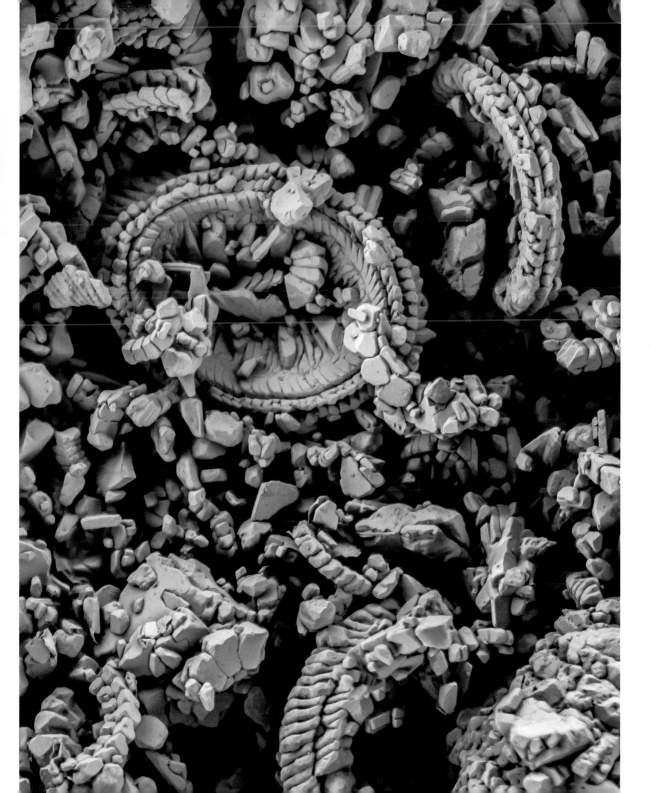

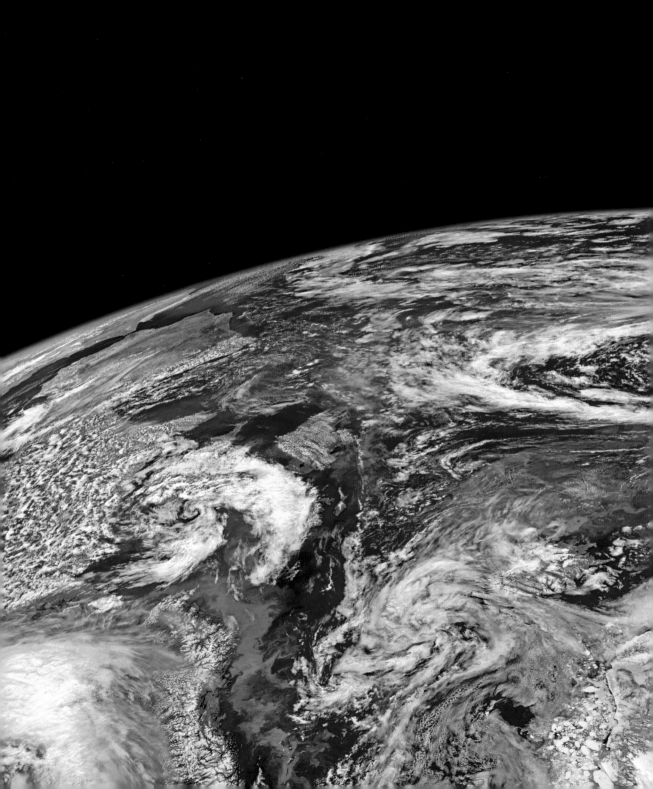

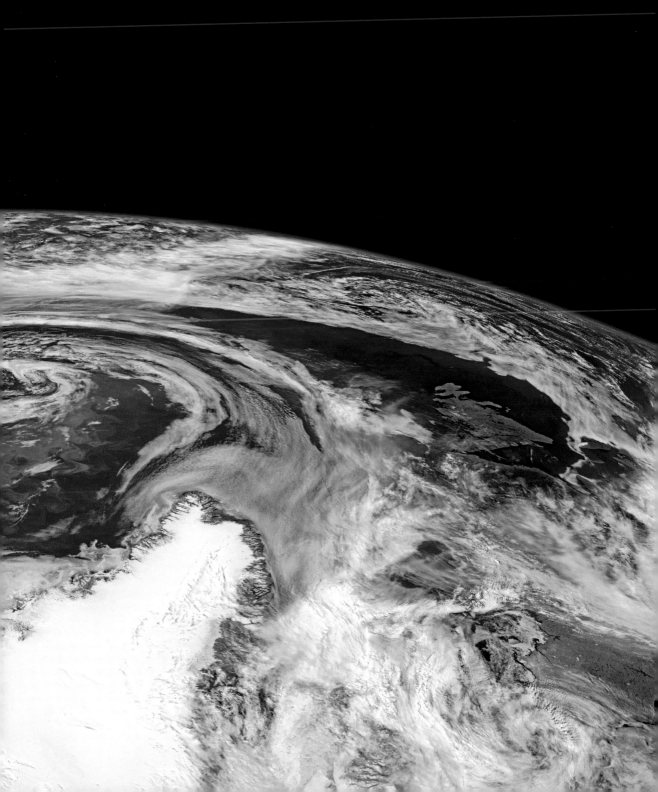

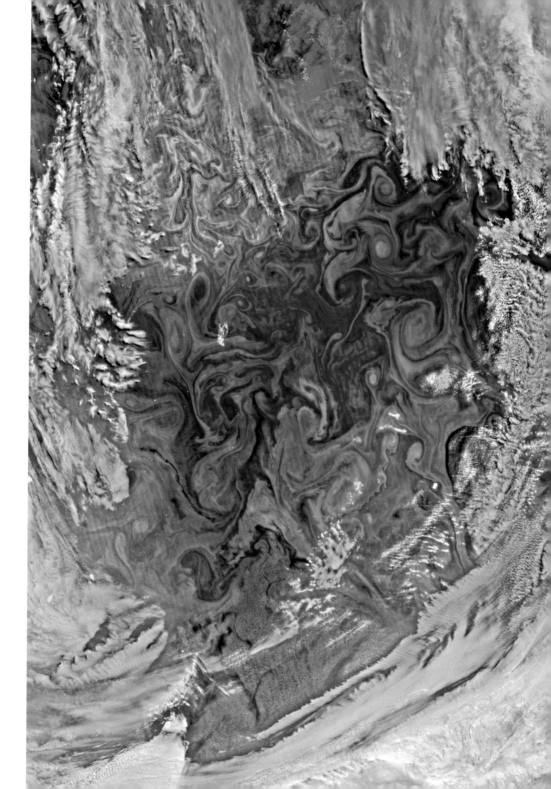

Brilliant patterns
swirling in the South
Atlantic Ocean.

signs of life on Earth is produced by creatures so small that they are invisible to the human eye?

It's a phenomenon that underpins nature as we know it. Look out to the horizon at sunset, and a single starling is undetectable against the bright sky. Yet ten thousand starlings in a flock instantly capture attention as they rise and tumble, as if they are one cohesive superorganism. Fish swimming in schools, lights in a city at night, and plankton in the ocean all become visible at great distances when clustered in groups.

Algal blooms emerge for months at time when and where conditions are right for trillions of individual phytoplankton cells to grow and coalesce. Born from the bottomless biodiversity of the ocean, an environment where millions of different microbes live, these fluid masses are composed mostly of cells from a single microbial species. The blooms we see were formed by a particular type of microalgae that has a unique microbial exoskeleton. The exoskeleton reflects light—so much light that the blooms are easily recorded by cameras mounted on NASA satellites. The exoskeleton is built of calcium carbonate plates that are made inside of the cell and exported out whole. If you are wondering what the exoskeleton is for, that's a good question. Microbiologists have theories, but they have yet to be confirmed. The exoskeleton might function as armor to protect the cell from predators and ultraviolet radiation, or perhaps it may serve as a shield against the ever-present predators of living cells, the viruses. It might serve as ballast for sinking into nutrient-dense water, or enable more efficient photosynthesis.

It is at this intersection in the story, before any evidence of broken discs seen in chalk with an electron microscope or phytoplankton blooms in the ocean seen from space, before any knowledge of plankton's reflective calcium exoskeleton, that we will move into the mind of another scientist on the brink of solving the very same mystery.

Thomas Henry Huxley lived and worked in England in the nineteenth century and has been known ever since for his nickname, Darwin's Bulldog. He earned this nickname through vigorous support for the then newly released theory of natural selection proposed by Charles Darwin. As a paleontologist and comparative anatomist, Huxley spent most of his time looking for relationships between living animals and fossils and comparing fossils from many different locations around the world. Around this period, the Atlantic Telegraph Company

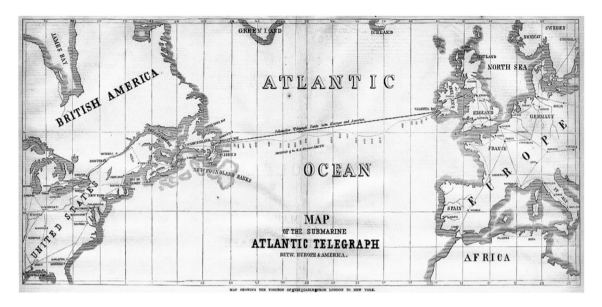

MAP SHOWING THE POSITION OF THE CABLE FROM LONDON TO NEW YORK.

The route of the first transatlantic telegraph cable between Ireland and Newfoundland, which opened rapid communication between North America and Europe in 1858.

was planning the first transatlantic telegraph cable between Europe and America. By chance, they hired an old friend of Huxley's, Lieutenant Commander Joseph Dayman of the Royal Navy, to join an expedition across the Atlantic Ocean. Dayman conducted depth soundings aboard the HMS *Cyclops* and brought up mud samples from the seafloor. This helped the company decide where to lay the cable. But to Huxley, it was an opportunity filled with wonder, a chance to explore an environment that was just beginning to see the light of science. He wrote that the newfound ability to scoop up mud from so far below the surface of the ocean "might have sounded very much like one of the impossible things which the young prince in the fairy tales is ordered to do before he can obtain the hand of the princess."

Captain Dayman's survey found a colossal plain along the Atlantic cable route filled with a fine, white mud. The mud looked and felt like soft chalk. Huxley analyzed the material and noted that if it is dried, "You can write with this on a blackboard, if you are so inclined." Working in the laboratory, he was shocked to find that most of the microscopic particles within the mud had a uniform size and an elaborate, round shape. He termed them "coccoliths," meaning "spherical rocks." Other scientists found coccoliths in Atlantic mud arranged

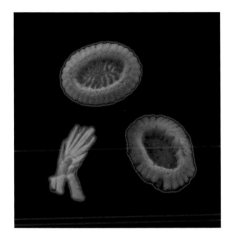

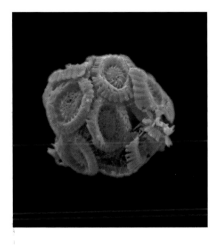

Round coccolith structures that Thomas Henry Huxley called "spherical rocks."

The coccolithophore *Emiliania huxleyi.*

into spheres and called them "coccospheres." Meanwhile, a man named Henry Clifton Sorby found coccospheres not within wet mud but in thin sections of dry English chalk. And unlike Huxley, who initially thought of these spheres as inert minerals, Sorby proposed that coccospheres are formed by living organisms.

Sorby was right. The coccoliths in the deep-sea mud and in chalk are the same calcium carbonate plates that reflect light and make living phytoplankton visible from space. This type of microbe, along with its coccolith exoskeleton, is now known as a coccolithophore. There are many species of coccolithophores, with a huge variety in shape, size, and layering of coccoliths. Some species carry as few as six coccoliths per cell while others have hundreds. The dominant coccolithophore in modern oceans is *Emiliania huxleyi*, a living memorial to Thomas Huxley. An image of *Emiliania huxleyi* from a scanning electron microscope shows its intact plated cell. The coccolith plates around the cell are continuously refreshed, with old plates falling off as new ones form.

Altogether, the dense mixture of plated cells and free-floating plates creates an aquamarine haze, wrapped up by ocean currents and contrasted against the blue water. Reflected light shows the exact size and location of coccolithophore blooms and other phytoplankton blooms. Or, we can measure the distribution of the pigment chlorophyll across the oceans of the globe at large, appearing red, green, and yellow where most concentrated, and light to dark blue where least concentrated. Blooms are one of many places on Earth where microscopic life becomes visible.

This composite image represents a full season of ocean chlorophyll concentrations in a range from 0.01 milligrams per cubic milliliter (mg / ml³; purple) through 5 mg / ml³ (red).

A phytoplankton bloom in the North Sea between England and the Netherlands. Antoni van Leeuwenhoek traveled across this area by boat in 1668.

There must have been coccolithophore blooms around England while Antoni van Leeuwenhoek collected his chalk samples from the beach in 1668. In fact, based on present-day satellite photographs of the North Sea between England and the Netherlands, the ship on which he sailed back to Holland might have passed right through one. Let's suppose it did.

As Antoni stood on the ship, millions of living coccolithophores and ejected, free-floating coccoliths fluttered within the water beneath him, descending one by one toward the ocean floor. The coccoliths slowly accumulated into deep-sea mud—the mud that Huxley studied in the mid-1800s. Where on the ocean floor was the single remote patch in that mud, amid the sunken calcareous ooze, where each phytoplankton particle that Antoni saw came to rest for eternity after its somber fall? Think of the eons of compaction, particles buried by relentless marine snow, then eons of churning within Earth's crust, cliffs rising millions of years later on a tectonic plate, eroding in the wind on a beach. All this took place before the ooze became the chalk in Antoni's pocket. The intricate discs and fragments in the ancient chalk graveyard are what remains of coccoliths made by living coccolithophores during the Cretaceous period. They lived long before Antoni's and indeed our own human ancestors evolved—before any primate lived on Earth. But as Antoni's ship came into port in south Holland and he returned to his home in Delft, he had no concept of the coccolithophore blooms, no way of knowing that the particles in the chalk were once living microbes in the sea.

Back in Delft, Antoni observed samples as quickly as he could collect them. He and he alone had a window into the microbial world. It was an astonishing accomplishment, especially considering his surroundings and the fact that he wasn't a scientist by training. Delft is a small, quaint city that rose from a rural village in the Middle Ages. Its history is marked by two gothic churches that tower above all other buildings, then and now: the Oude Kerk and the Nieuwe Kerk. The seventeenth-century Delft where Antoni lived is captured in maps of the city from the time and the paintings of another famous Delft resident, Johannes Vermeer. Antoni sold clothing for much of his life. Had he not owned a clothing shop, he never would have started making microscopes. He intended to use a microscope to assess the quality of fibers in the fabrics he sold, which was a common practice among clothing merchants. He just happened to be exceptionally good at making the magnifying lenses—so good that he discovered microbes by accident.

Antoni constructed hundreds of microscopes throughout his lifetime. Some were aquatic microscopes for examining water from local ponds and canals; others had pins for dry samples like the one he used to observe chalk. He set up infusion experiments in pepper jars, gathered rainwater from his roof, and looked at samples from his own body. Along the way, he was the first to document the existence of bacteria, red blood cells, yeast, and sperm cells. Among the most well-known sketches of what he saw through his microscopes are the first images of bacterial cells. A famous illustration, dated September 17, 1683, shows rod-shaped bacteria now called bacilli and round cells, or cocci, scraped from the plaque on his teeth. Antoni called all of these organisms "animalcules." He watched the "many very little living animalcules" on his teeth "very prettily a-moving." Some of these animalcules, he wrote, "had a very strong and swift motion, and shot through the water (or spittle) like a pike does through the water." His only available comparisons at the time were with macroscopic animals. He had no other resources, for most of the microbiology in the late seventeenth century consisted of the observations he sent off in letters to the Royal Society in London.

Delft, the Netherlands, as it was during Antoni van Leeuwenhoek's life.

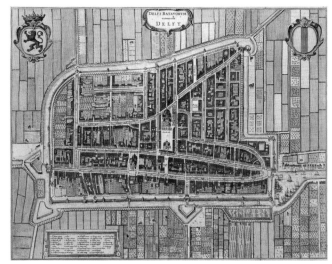

View of Delft, Johannes Vermeer, 1660–1661. The painting shows the main harbor leading into the canals of Delft.

The microbial universe that Antoni discovered would one day revolutionize our understanding of the natural world. But in the late 1600s, many were skeptical of Antoni's observations. Robert Hooke was an English scientist working at the Royal Society when Antoni's first letters arrived. Hooke was one of the handful of people on Earth making microscopes at the time. Today he is recognized for first using the word "cell" and for his 1665 masterpiece, *Micrographia*. Bacteria and other microbes were doubtlessly present in many of the samples that Hooke observed and wrote about in *Micrographia,* but his microscopes were not powerful enough to visualize them. Eventually, Hooke saw the animalcules for himself, verifying Antoni's reports. By the end of Antoni's life, dignitaries from across the world visited his house in Delft for a peek through his lenses into an otherwise invisible layer of life. Antoni observed more microbes than anyone else would for over a hundred years following his death, in part because nobody knew precisely how he made his lenses.

Now we use the term *microbe* instead of animalcule. We can identify microbial species and place them in a universal "tree of life." By doing so, we understand their evolutionary relationships to other organisms. Trees of life have been a part of many mythologies since ancient times, but it was Thomas Huxley's friend Charles Darwin who famously extended the concept into an evolutionary context. Darwin wrote, "As buds give rise by growth to fresh buds, and these, if vigorous, branch out and overtop on all sides many a feebler branch, so by generation I believe it has been with the great Tree of Life, which fills with its dead and broken branches the crust of the earth, and covers the surface with its ever-branching and beautiful ramifications." A diagram of a tree with branches that symbolize evolutionary relationships between biological entities was the only illustration in Darwin's 1859 masterwork *On the Origin of Species*. It does not explicitly show relationships for any one group of organisms. Rather, it shows a tree as an abstract concept: a theory for how one might depict divergence from common ancestors and evolutionary change over time. After Darwin's rendition of the tree of life, naturalists continued drawing phylogenetic trees based on comparisons between the physical features of organisms. Making sense of the tremendous diversity of life by comparing physical features alone turned out to be an impossible task, however, particularly for the microbes.

If we want to identify the many different microbial species that live someplace like the human mouth, the physical appearance or morphology of cells is

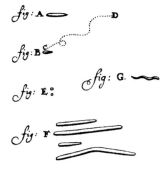

Bacterial cells as seen through Antoni van Leeuwenhoek's microscope in 1683. This illustration was sent in letter no. 39 to the Royal Society in London.

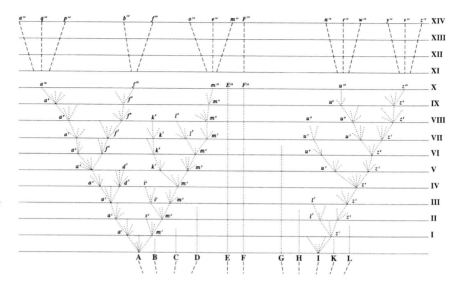

The tree diagram from Darwin's *On the Origin of Species*, 1859. The horizontal lines are generations. The labels and numbered letters represent distinct species and subspecies.

not enough. The smallest genetic change could cause two cells from the same bacterial species to look different. Or two species that might be only distantly related to each other, more genetically distinct than an ocelot is to an orangutan among mammals, might look exactly the same.

The advent of molecular biology in the 1950s completely changed how we draw trees of life. Its central dogma relied on genetics rather than appearances: DNA sequences called genes are transcribed into RNA sequences, which are translated into proteins. Soon after, American scientists Carl Woese and George Fox were among the first to use genetic information instead of physical features to determine evolutionary relationships between organisms. They chose the gene for their study carefully; it had to be a gene that encodes a function so important for the most basic operation of the cell that it is found in all forms of life, from microbes to grizzly bears. They chose a gene named "small subunit ribosomal RNA," which encodes a piece of the ribosome, the cellular machine that converts the genetic information that is transcribed into RNA into functional proteins. With one scientific paper in 1977 and another from Woese in 1990, their work on ribosomal RNA genes reshaped the tree of life from a five-kingdom system to just three major domains: Archaea, Bacteria, and Eukarya. Archaea were a completely new group of microbes in 1977, joining bacteria and eukaryotes that

The three-domain tree of life represented in a metal sculpture: Bacteria, *upper left*; Archaea, *upper right*; Eukarya, *below*.

are composed of more complex cells, like microscopic eukaryotes, fungi, slime molds, plants, and animals. The version of the tree of life showing the three domains as a metal sculpture is based on one described by the American microbiologist Norman Pace in 1997.

While Woese and Fox were exploring genetic relationships between organisms in the 1970s, the American scientist Lynn Margulis was equally fast at work formulating a new theory: the theory of endosymbiosis. By this time, scientists were imaging organelles and other structures inside of cells with electron microscopes. Based on electron microscopy and other evidence, in a landmark paper published in 1967 Margulis proposed that complex eukaryotic cells arose through a merging of several previously free-living organisms. Margulis theorized that the mitochondria that produce energy within eukaryotic cells were once free-living bacteria and that chloroplasts which harvest light energy in plant and algae cells were once free-living photosynthetic bacteria, the cyanobacteria. In the decades that followed, these cellular organelles were found to have their own genomes and other vestiges of their previous lives as free-living organisms. Margulis's once radical idea became a central tenet of biology.

The microbial species that merged with a bacterium and became the last universal common ancestor of all eukaryotic cells no longer exists exactly as it did billions of years ago when the event is thought to have occurred. In fact, as the gears of evolution grind over the millennia, most species that ever lived on Earth are now extinct. The best that we can do to understand more about the origin of eukaryotic cells is to find microbes that have properties similar to those of the first eukaryotic cell.

There is one archaeal species at the bottom of the ocean that has many of the traits we would expect of the host cell that merged with a bacterium to become the first eukaryote. The microbe is an archaeon whose genome was found near a hydrothermal vent called Loki's Castle. That's how it got the name Loki-archaeota, or Loki for short (after the Norse god of shape-shifting). Loki was found using an approach called metagenomics, through which all of the genetic material in a given environment is sequenced and reassembled into separate genomes after the fact. This means that though we know Loki exists and though we have many insights into its biology via its genome, we have yet to trap any living Loki cells. Loki's genome indicates that this organism is closely related to modern eukaryotic cells. For example, Loki has the genes for a primitive cyto-

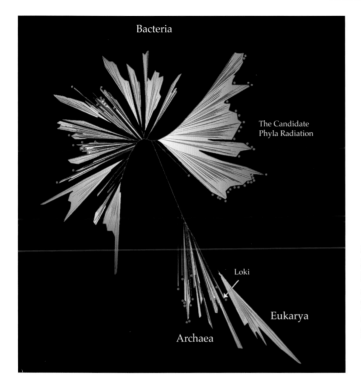

Bacteria

The Candidate
Phyla Radiation

Loki

Eukarya

Archaea

A new tree of life: Bacteria expands, and Eukarya groups within Archaea. The branches of this tree represent taxonomic groups known as phyla. The relationships between phyla are based on a combination of sixteen separate ribosomal protein sequences. The tree captures ninety-two named bacterial phyla, twenty-six archaeal phyla, and five supergroups of eukaryotes, numbering 3,083 different organisms in total.

skeleton, an internal scaffolding that maintains and changes the shape of the cell and organizes cellular components. Such a cellular remodeling system would have allowed a species like Loki to engulf large objects from the environment, such as another microbe. Loki is joined by several other archaeal groups with eukaryotic traits named after Norse myths, like Thorarchaea and Odinarchaea, together called the Asgard archaea.

Evolution is multidimensional. Woese, Fox, and Pace gave us the basic shape of the tree of life and separated the Bacteria from the Archaea, and Margulis showed us that the branches of that tree are malleable: cells within many organisms can be broken into several parts, each with its own evolutionary history. Our understanding of the tree of life changes over time, often rapidly.

Skipping ahead to the newest version of the tree, we find there is now an entirely new group of bacteria called the candidate phyla radiation. The discovery of this bacterial group nearly doubled the total known biodiversity on Earth.

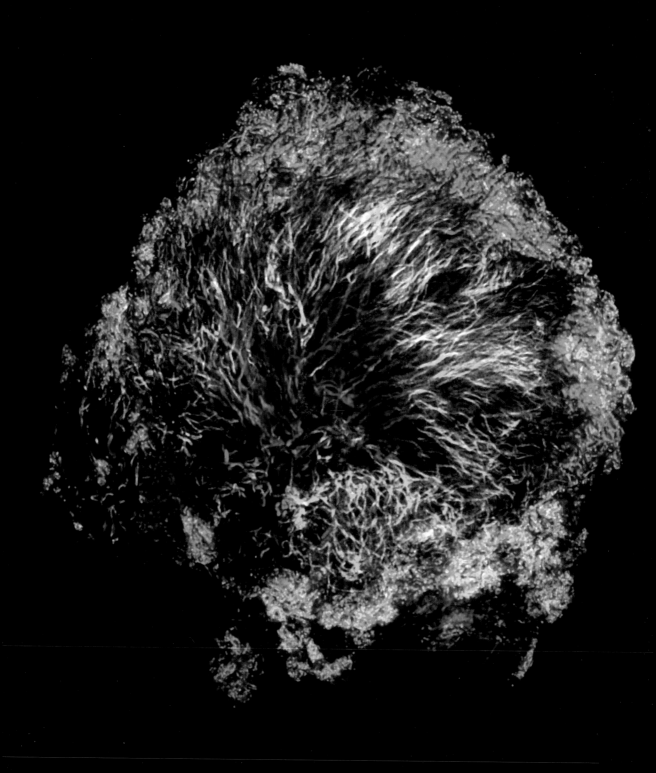

Members of the candidate phyla radiation are found in virtually all microbial ecosystems—places known as microbiomes. Most of these bacteria cannot be studied in laboratory conditions. Sometimes known as ultramicrobacteria, they are extremely small, and many are likely to be ecto- or endosymbionts—species that live attached to the outside or inside of other species and are dependent on them for survival. Before long, we will learn how to work with these finicky microbes, either by growing them together with the other species within their local communities, or by figuring out exactly what they are getting from their symbiotic partners and supplying this within their growth medium.

As they did when they were discovered by Woese in 1977, the archaea are again reshaping the tree of life. The genomes of more and more archaeal species were sequenced and used to make new phylogenetic trees. Eventually, the entire domain Eukarya was clustered within the archaeal part of the tree rather than out on its own branch. Eukaryotes emerged from within the Archaea. So while eukaryotes have special qualities and certainly deserve their own name, every eukaryotic cell can be viewed as an archaeal cell. All of our human cells contain ancient artifacts of the bacterium that became the mitochondrion.

Not only can we use ribosomal RNA sequences to name the microbes in our mouths, we can also use them to understand how each species lives in community with others. We can tease apart many kinds of microbes to see these organized communities. With fluorescence microscopy we can label many different species in dental plaque and other environments simultaneously, each as a different color, tracking the locations and interactions of individual cells. The resulting multicolored structures reveal a level of sophistication in the formation of these microbial communities we could not have imaged just a few years ago.

Not only communities but individual archaea and bacteria in the mouth are now visible with these new tools. From there, we zoom in even closer to see minute structures that enable cells to go about their lives. In other words, we can now ask this question: What are the microbial cells doing? The archaeal and bacterial cells are surrounded by a cellular envelope. Bacteria either have a single cell membrane composed of phospholipids surrounded by a thick cell wall, in which case they are known as gram positive; or they have two membranes that sandwich a thin cell wall, in which case they are known as gram negative. These two major types of bacteria are based on the Gram staining method used to differentiate cells under the microscope, invented by Hans Christian Gram in the

The hedgehog microbiome from human dental plaque, seen with fluorescence microscopy from the lab of Gary Borisy. Bacterial groups shown: *Streptococci* (green), *Corynebacterium* (pink), *Leptotrichia* (light blue), *Capnocytophaga* (red), *Porphyromonas* (blue), *Neisseriaceae* (purple), and *Haemophilus* (orange).

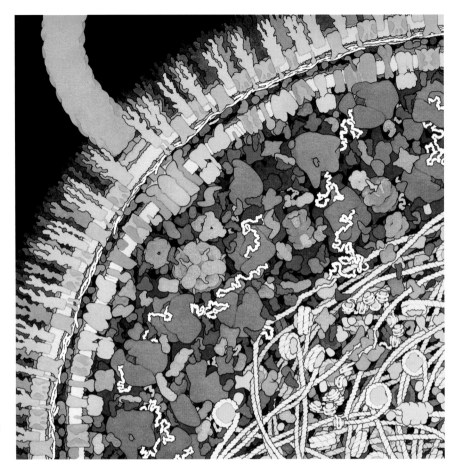

Molecular biologist David Goodsell's watercolor rendition of the complex and crowded realm of molecules and molecular machines inside of a swimming *Escherichia coli* cell. The green molecules show the cell membranes and cell wall, as well as a green flagellum extending from the cell with its flagellar motor complex at the base. Transmembrane proteins are also shown jutting through the cell membrane. The cytoplasmic region is shown as purple and blue molecules. The large purple molecules in the cytoplasm are ribosomes, the white strands are mRNAs, and the smaller, maroon molecules are transfer RNAs. Finally, genomic DNA is shown in yellow, wrapping around bacterial nucleosomes.

late 1800s. Archaea, like the gram-positive bacteria, have one cell membrane, but the archaeal membrane is composed of different lipids than those in the bacterial membrane. Archaeal cell envelopes often also have an outer crystalline lattice of proteins called an S-layer.

Gram positive, gram negative, or otherwise, the cell envelopes of microbes are innervated by membrane channels that bring different molecules into the cell or pump molecules out into the environment. They are also decorated by an array of peripheral proteins that are attached to the cell where they perform specific functions, like binding or breaking down molecules from the environment

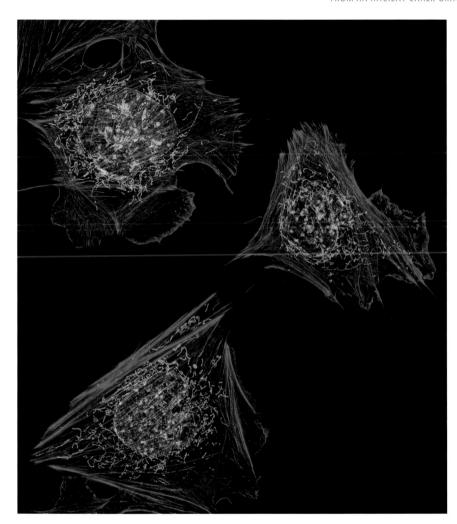

The eukaryotic cell. Fluorescence confocal microscopy of mouse fibroblast cells (connective tissue). Mitochondria are represented in green, the nucleus is blue, and the cytoskeleton is red.

for food. Many bacteria can swim through the environment using corkscrew-like cellular appendages called flagella. Archaea have their own appendage for swimming that evolved independently from the bacterial flagella; it's called the archaellum.

Eukaryotic microbes can be free-living individual cells with complex internal compartments or animals that just happen to be too small to see. Eukaryotic

The red rooftops of Delft today, as seen from the top of the Nieuwe Kerk, with the Oude Kerk in the distance.

The canal at the site of Antoni van Leeuwenhoek's house in Delft. A small plaque commemorates Leeuwenhoek on the column in the center.

A miniature ecosystem grows at the edge of the canal.

cells are almost always larger than bacterial and archaeal cells, and they all share a set of key features not seen in bacteria or archaea. Genetic information is organized into chromosomes that are stored within a membrane-enclosed cell nucleus. Cellular energy is produced within mitochondria. Proteins and other cellular cargo are distributed through a network of flattened membrane disks, including the endoplasmic reticulum and the Golgi apparatus.

All told, over billions of years of evolution in bacterial, archaeal, and eukaryotic organisms, how many microbial species are on Earth? It will be a very long time before we can identify each and every microbe in every habitable microbiome, so the best way to estimate the total number of species is to model microbial biodiversity mathematically, scaling up from all of the microbiomes that we have analyzed so far. These models suggest there could be as many as one trillion different species of microbes.

This modern view of microbes is the product of many scientific disciplines, together known as the microbial sciences, that are based in or intersect with the microbial world. From this perspective, large animals are the oddities in a world teeming with microscopic life—not the other way around. Right now under the red rooftops of Delft, in one gram of soil where Antoni van Leeuwenhoek's home once stood, there are tens of thousands of microbial species. Almost all of these microbes lack scientific names. We are just beginning to learn how to isolate and study them. It's a frontier in biology.

The microbes living across the planet are the foundation of the entire global ecosystem. Prehistoric phytoplankton fossilized in natural chalk and those in the ocean today have produced most of the oxygen on Earth—the same element that makes up 65 percent of the human body by mass. Much of the energy driving the metabolism in every cell in our bodies right now was first harvested from the Sun by phytoplankton. This energy then cycled up through zooplankton, which ate the phytoplankton, and then passed through larger creatures that ate those zooplankton—creatures we call seafood.

Without microbes, human society could not function. Microbes make valuable medicines, filter our waste water, and clean pollution. They make many of the most delicious foods we eat, including chocolate, wine, bread, and cheese. The characteristic flavors of aged cheeses develop from ecosystems of bacterial and fungal species, safely eaten with every bite. One after the other, genes that evolved inside microbial cells are brought into the spotlight to become advanced human biotechnologies. How many technologies remain undiscovered in a microbe, whether it lies on a piece of Antarctic ice, at the bottom of a boiling Yellowstone mud pot, or on your skin? Many of these may revolutionize human society in the future. In addition to aiding humankind, the microbes will always be fascinating life-forms.

Microbes open a window into the nature of life itself. They are grown, manipulated, and imaged in the laboratory with unmatched ease and accuracy. Microbes allow us to address the most fundamental questions in biology. What was the origin of life? Is there life on other planets? How does evolution occur? How do cells organize? How do ecosystems function? How do unlike species change and merge through symbiosis? How did multicellular organisms come about? Or even, what is life?

The microbial world is a beautiful place to explore. There is as much biodiversity and behavior to be studied on the tiny fronds of duckweed floating in a canal in Delft as there is in an Amazonian rainforest. Microbes form colonies

The mud volcanos of Yellowstone National Park, Wyoming, where thermophilic microbes like archaea from the genus *Sulfolobus* live. Tourists watch Dragon's Mouth Spring in the background.

The canals of Delft filled with duckweed. Antoni van Leeuwenhoek is buried beneath the leaning Oude Kerk in the background.

Every frond of duckweed is a
microbial ecosystem.

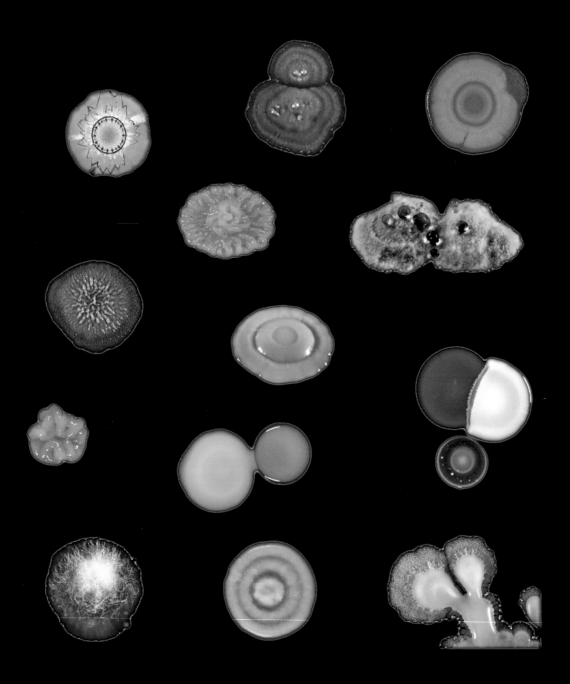

containing millions of cells that grow to the size of a coin or larger, with exquisite structure and shape, producing pigments that represent all of the colors on an artist's palette.

This is the microbial world we endeavor to show within the following chapters. It's a world that is both ancient and ever changing, buried in billion-year-old fossils and driven by constant cellular reactions operating on the scale of femtoseconds. It's a world based in ecosystems and food webs, where microbes communicate, cooperate, and compete with one another, building communities and engaging in collective behaviors, where they network with larger organisms, evolving faster than any other kind of life, right on our own bodies and kitchen tables. Biology is in renaissance, and the renaissance is rising up from the micro-bial world.

A variety of bacterial and fungal species form macroscopic colonies, each about the size of a dime. Different species shown produce natural orange, yellow, and red pigments.

2

To the Heartbeat of Earth

ON A SUMMER NIGHT IN YELLOWSTONE NATIONAL PARK in Wyoming, a young man carefully stands on a wooden boardwalk, shining his headlamp into the distance. He sees a landscape unlike any he has ever imagined: bright orange streaks radiate across a flat, cracked surface covered by hazy steam clouds. The air is warm with the smell of sulfur. He is at the largest hot spring in North America: the Grand Prismatic Spring. It is over a hundred meters wide and filled with boiling hot water fifty meters deep. Alone under the stars, he imagines this place could be from another world. The next day, the visitor hikes up a trail above Grand Prismatic Spring. Only from this vantage point can he appreciate the scale of the patterns and colors he saw the night before. He watches the many tourists far below, filing across the boardwalk. He sees Excelsior Geyser Crater next to Grand Prismatic, a baby-blue chasm in the landscape that dumps its boiling water into the Firehole River.

Geothermal phenomena in Yellowstone are telltale signs of a hidden supervolcano. Cool water from the surface flows through underground aquifers and slowly rises in temperature thanks to two gigantic magma chambers in this area of Earth's crust. From the aquifers it returns to the surface as superheated water and steam. Heat energy in these magma chambers comes from an eighty-kilometer-wide plume of molten rock: a hot spot within Earth's mantle. At the base of the plume is the core of the planet itself, parts of which are hotter than the surface of the Sun.

Where did this energy come from? Most of the heat in Earth's core today is the result of the radioactive decay of uranium and thorium. There is also some primordial heat—energy left over from when the planet first formed. So when we see a geyser erupt on Earth's surface in a place like Yellowstone, when we feel giant pockets of air burst through hot springs, causing the ground itself to pulse rhythmically, we are in some small way experiencing the lingering energy of events that took place in the Solar System 4.5 billion years ago. It is a way of sensing the enduring heartbeat of Earth.

Grand Prismatic Spring, Yellowstone National Park. A shooting star is seen in the sky from the Perseid meteor shower.

Grand Prismatic Spring and Excelsior Geyser Crater, Yellowstone National Park.

Tourists walk over the orange microbial mats at Grand Prismatic Spring. The yellow innermost band of microbial mat within the spring is the hottest region (other than the blue center) and contains *Synechococcus* cyanobacteria. *Synechococcus* remain in the cooler, orange region of the mat and are joined by many other thermophilic microbes, including *Calothrix*, *Deinococcus / Thermus*, *Phormidium*, and *Chloroflexus*.

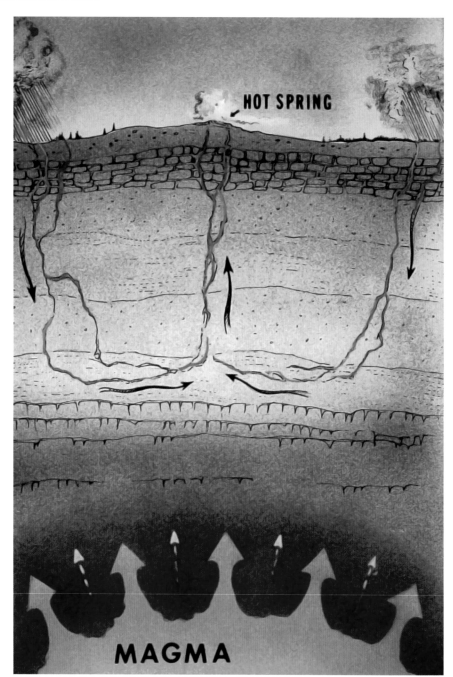

Geothermal features in Yellowstone are caused by a hidden supervolcano. This vintage diagram from Yellowstone National Park shows the natural forces that drive hot springs.

That's when Earth took shape within a cloud of interstellar dust and molecules that collapsed under gravity to form the Sun. We know this by watching other examples of these solar nebulas in far-flung regions of the cosmos using space telescopes. Whatever was not enveloped within the Sun was cast around it in a protoplanetary disc. Objects ranging in size from pebbles to giant planetesimals crashed and gathered within the disc, aggregating into larger and larger objects. Earth was formed from this celestial violence. The composition of the planet, the uranium and other radioactive elements producing the heat in Earth's core, was set in stone during those random crashes. Shrapnel from

Stars like the Sun form from clouds of interstellar gas and dust. Here an area of cold interstellar gas can be seen within the Carina Nebula, 7,500 light-years from Earth in the constellation Carina. Oxygen is seen as blue, nitrogen and hydrogen as green, and sulfur as red.

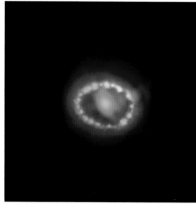

The supernova SN 1987.

Earth's planetary formation still floats in the asteroid belt between Mars and Jupiter.

If we track the energy yet another step back in time, we find that all of the atoms that formed the planets and asteroids in our Solar System were produced inside a dying star and thrown into the universe in a titanic supernova explosion. Supernovas are constantly occurring somewhere in the universe, giving rise to new atoms. Within our own galaxy, the Milky Way, there is at least one supernova every century. If we could monitor the universe as a whole, a few hundred stars exploded somewhere in the time it took you to read this sentence.

Go outside wherever you are and pick up an ordinary rock. From how many different locations within the universe did the atoms in that rock come? From which directions? From exactly how many supernova explosions? Precise answers to those questions are unattainable with the science of today. The same questions can be asked about the atoms held within the photoreceptor proteins in your eyes—the ones you use to see this text.

Take the example of SN 1987A, a supernova named for its light, which reached Earth in 1987. SN 1987A is located in the outskirts of the Large Magellanic Cloud, a dwarf galaxy 163,000 light-years away. It's so far away that when the light first left the supernova 163,000 years ago, it was the Middle Paleolithic period on Earth. Humans were just learning to wear clothing.

The brightest supernova before SN 1987A occurred before the invention of the telescope, just 20,000 light-years away from an exploding star within the Milky Way. That one was so bright that it was visible to the naked eye. It was first spotted from northern Italy on October 9, 1604, and on October 17, the astronomer Johannes Kepler viewed it from Prague and continued to observe it for over a year. It was still visible in the sky on November 1 when William Shakespeare's *Othello* was performed for the first time in London. The matter within those bright shockwaves spread and cooled. Will the cosmic dust collect somewhere to form an asteroid, wrapped up within a giant mass next to another sun, in a world like Earth?

When Earth first formed, it was nothing like the green and blue planet we know. It resembled what many traditions envision as hell. It was the Hadean eon, named after Hades, the ancient Greek god of the underworld. The planet was an ocean of magma. Earth was so hot and fluid that the densest elements sank toward the center of gravity, forming a core made almost completely of pure iron and nickel during a planetary phase known as the iron catastrophe. Lighter elements, including carbon and hydrogen, were left floating at the surface.

Even though the formation of the molten metal core of Earth was a purely physical phenomenon that occurred at 2,000 degrees Celsius (°C), it was an essential step toward the evolution of life. As the early Earth spun on its axis, differences in temperature and pressure within the liquid metal core created convection currents and generated electric current. It was like flipping a switch that turned on a magnetic field around the entire planet: a field known as the magnetosphere. Without the magnetosphere, the continuous stream of charged particles emanating from the Sun would have prevented life as we know it from

The magnetosphere shields Earth from the ever-present solar wind. A corona mass ejection emanates from the surface of the Sun, causing a blast of solar wind that is mostly blocked by Earth's magnetosphere.

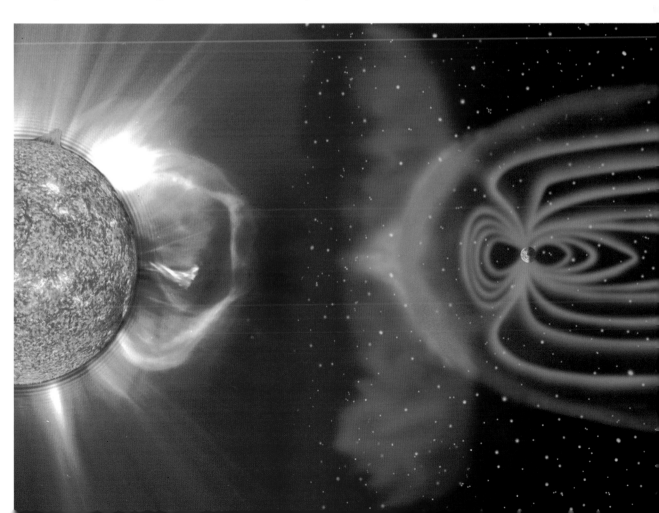

arising. The magnetosphere acts like a shield; if it somehow powered down on today's Earth, the solar wind would instantly disable all satellites and electricity grids. Radiation from cosmic rays would start the timer on a worldwide experiment, increasing the mutation rate of DNA in millions of organisms at once. The atmosphere, the delicate sheen on the surface of Earth that forms the physical framework of our biosphere, would blow away into space. We don't have to speculate about the results. We are very much aware of what planets look like after their atmosphere has been stripped away by solar wind. This happened to Mars. If it all seems too abstract, keep in mind that it is the magnetosphere that aligns the metal needles in our compasses. Many organisms, from magnetotactic bacteria to migratory birds and mammals, can sense magnetic fields and navigate according to them. There are also places you can go right now to see the solar wind bashing against the magnetosphere. This is what causes aurora borealis and aurora australis, the northern and southern lights.

Once Earth was protected from the solar wind, emerging forms of life had to contend with another few hundred million years of heavy bombardment by asteroids. Between 4.1 to 3.8 billion years ago, meteorites the size of the one that later caused the mass extinction of dinosaurs were hitting Earth regularly. Using the cratered surfaces of the Moon and other rocky planets in our Solar System as a guide, the total mass delivered to Earth from space during this time is estimated at 220 billion teragrams. That's the equivalent of 4 trillion cruise ships, or 36 billion times more mass than all of the blocks in the Great Pyramid of Giza in Egypt. The heat generated by these impacts, residual heat from planetary formation, volcanic activity, mixing of the planet's crust through tectonic activity, and billions of years of erosion have destroyed rocks from the Hadean eon in all but a few places on Earth.

The Acasta Gneiss in the Northwest Territories of Canada and the Jack Hills in Western Australia are two of the oldest rock formations on Earth. Rocks contain zircon crystals that act like date markers. Each tiny crystal forms with a certain amount of trace uranium. By measuring how much uranium transformed into lead over time through radioactive decay, we can estimate the Acasta Gneiss in Canada as 3.96 billion years old, making it the oldest intact fragment of continental crust on Earth (the stuff that we call "land"). Zircon crystals are so durable that they can push our view of Earth's geological history past the earliest rocks. In the Jack Hills of Australia, there are zircon crystals dated at 4.374

Mars, without its
magnetosphere.

An aurora glows green as solar
wind interacts with Earth's magnetosphere. In this image of the skies
over Europe, sunrise appears blue,
close to the horizon.

billion years old, plus or minus 6 million years. These Australian zircon crystals are the oldest terrestrial material known—older than the rock matrix that they are deposited within. The zircons reveal that continental crust on Earth formed in the Hadean eon, not long after the Earth itself formed.

The version of life shared by every cell today can be traced through an unbroken chain of cellular divisions to some time and place within this Hadean world. Meteor impacts were common, but they did not produce enough heat to prevent life within the Earth's lithosphere (the region of Earth that includes the crust and upper mantle). The limits of life on Earth today range from heat-loving or thermophilic archaea that survive temperatures of 121°C and cryophilic or cold-loving bacteria that persist in 20°C permafrost. The latest simulations indicate that even if all the meteorites and comets from the late Hadean eon came upon the planet simultaneously, temperatures still wouldn't have been high enough to melt all of Earth's early crust and eradicate up-and-coming life-forms. Sometime in the late Hadean eon, the molecules of life were being assembled into the first cells. And since these cells were likely too small to see with the naked eye, we can declare with confidence that the age of microbes was under way.

Envisioning the first microbial life on Earth requires a conceptual deconstruction of what it means to be a cell. All modern cells have a membrane made of lipids that separate an intracellular space from the extracellular space. They have a molecule capable of storing information within the cell: a molecule arranged as a sequence of nucleobase subunits that can self-replicate. This is the heritable material of life, which for most organisms today is DNA. The information is then transcribed into RNA and translated into functional molecules that form the structural pieces of the cell and enzymes that catalyze chemical reactions. The combined activity of these functional molecules, including catalytic RNA molecules and proteins, is the metabolism of the cell. Metabolism can be defined as all of the chemical reactions that occur as molecules are brought in from the environment, broken down, and used to synthesize other compounds needed to sustain the life of the cell. When we think of early life, we think of protocells, not modern cells. Imagine protocells as microscopic "spheres" containing a combination of primitive molecules. These primitive molecules satisfy the basic functional categories of a cell membrane, which are information storage and enzymatic or metabolic function. To trace one way that the first protocells might have arisen, we will follow several storylines that take place in parallel—

stories where the characters are the chemical precursors of life that will, in the end, intersect.

First is the chemical history of the membranes that must have defined the physical space of the first protocells. Far below the ground, minerals lining a wall inside of a geyser similar to those seen in Yellowstone today catalyzed the formation of simple types of lipid molecules called fatty acids from carbon monoxide and hydrogen. When the geyser erupted, the fatty acids were ejected into the air, floating off in the wind and landing in a pond. Through many eruptions like this, the fatty acids collected in the pond and became concentrated. One end of each fatty acid was attracted to water (hydrophilic), and the other end was repelled by water (hydrophobic). This single property of fatty acid molecules led to the spontaneous formation of micelles, spheres with the hydrophilic end of the fatty acids outside and the hydrophobic end inside. Micelles then merged to form a sheetlike bilayer, and the bilayer sheets elongated until random motion brought the two ends of the molecules together, forming spherical membrane vesicles. These basic membranes arose spontaneously; they self-assembled under the right conditions.

Meanwhile, a meteorite seared through the sky above Earth. As it entered the atmosphere and struck land, its kinetic energy added enough heat to synthesize hydrogen cyanide. The atoms needed for this reaction—carbon, nitrogen, and hydrogen—were already present in the atmosphere, in the early oceans, in the early crust of Earth, and within the asteroid itself. Hydrogen cyanide accumulated and participated in further chemical reactions driven by ultraviolet light from the Sun, producing precursors of nucleotides: the subunits of nucleic acids. The nucleotides traveled through streams and groundwater, ending up in the same pond where fatty acids were forming bilayer sheets and spherical membrane vesicles.

Then, an inanimate particle of clay in the pond became enveloped by a newly formed vesicle. The clay particle was next joined inside of the vesicle by some of those newly formed nucleotides. Nucleotides moved freely into the vesicle as a result of constant shifting among fatty acids in the membrane. Once inside, they were attracted to the charged surface of the clay particle. As the clay particle became covered by nucleotides, the nucleotides were forced into close physical proximity by their common attraction to the clay atoms. This catalyzed the formation of chemical bonds between the nucleotides, packed against each other on the clay. They formed short molecules of a single-stranded nucleic acid similar to modern RNA. These very first RNA polymers could in turn act as a template for another RNA molecule. In this secondary round of RNA polymerization, nucleotides were attracted not to the clay but to other nucleotides along existing RNA molecules, once again forming bonds with adjacent nucleotides as they lined up next to each other on the template RNAs. At this point, the RNAs were self-replicating the sequence of the first genetic molecule. They weren't replicating quickly or accurately. A modern RNA polymerase enzyme can zip out RNA molecules at fifty nucleotides per second, making a mistake once per 100,000 nucleotides or so. With a clay particle or a crude RNA template acting as a catalyst instead of RNA polymerase, synthesizing short RNA molecules may have taken days, in part because the synthesis relied on passive diffusion of nucleotides from the environment into the vesicle. Nevertheless, a primitive copying mechanism for a genetic material was at work.

Some of the RNA-filled vesicles were swept into a hot area of the pond. This change from a cool area that promoted vesicle formation and RNA polymerization to a hot environment separated the two strands of the RNA molecule. The heat also made membranes more permeable, driving a period of rapid integration of fatty acid micelles from the environment and therefore enhancing vesicle growth. Vesicles enlarged, and with a bit of mechanical energy in the form of a wave in the pond, they broke apart to produce daughter cells. These were the first cellular divisions.

Together, fatty acid vesicles, an RNA-like nucleic acid, and clay particles were primitive protocells. Instead of being composed of millions of different molecules in the manner of a modern cell, these cells may have been made of 10,000 molecules. RNA was the self-replicating information molecule. Vesicles were the compartments that separated the protocytoplasm from the environment. And the clay, like a primitive enzyme, was a hub where molecules interacted, catalyzing energetically unfavorable reactions from a mixture of inorganic compounds.

Continuous influx of fatty acids and nucleotides fed the environmentally driven growth cycle and led to large populations of protocells within the pond, each with a mixture of randomly synthesized RNAs. The longest RNAs spontaneously folded into three-dimensional shapes because of interactions between

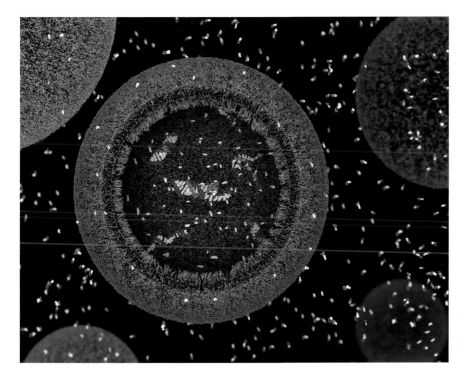

A protocell with a bilayer cell membrane and genetic material inside.

nucleotides in the molecule. By sheer chance, one of these folded RNA molecules fit around other RNA molecules in the protocell in such a way that it stabilized and sped up the rate of RNA replication. This rudimentary RNA replicase was a ribozyme; it was the first enzyme on Earth. RNA can both store information and catalyze chemical reactions. A basic RNA replicase, similar to what we think an early ribozyme would look like, can amplify other RNA molecules over 10,000 times within one laboratory experiment, including complex RNAs with cellular functions.

Darwinian evolution had begun in an RNA world. RNA encoded information in the form of nucleobase sequences, and in doing so, it acted as a kind of molecular memory. Those chemical sequences then spontaneously folded into molecular shapes and structures, which acted as primitive phenotypes: physical manifestations of genetic sequences that could interact with other molecules in the environment. And the sequences were replicating inaccurately, so that new variations were constantly arising in the RNA sequences. Some of these variations would be advantageous by chance, increasing the fitness and reproductive success of the protocell lineage that they were contained within.

It was an era of competition and cooperation between protocells. One protocell lineage may have developed a set of competitive advantages, yet all cells were continuously trading fatty acids between their membranes. There were no specialized membrane pores or active transporters for regulating what entered and exited. Protocells were leaky. RNAs circulated between different cell lineages. Molecules within each cell were randomly assorted between daughter cells during cell division. There was a mechanism for passing down advantageous traits: the basis for heredity. Individual cells had identities. But the identities were weak. Any given molecular innovation arose in a given protocell, replicated inside of it, and then mixed between cells to spread within the population.

During this early period of cell history, one cell lineage became the last universal common ancestor of all living cells today, nicknamed LUCA. Through a series of intermediate steps that we do not yet understand, LUCA had already moved on from the RNA world. Information was stored in DNA instead of RNA. Single-stranded RNA was used to translate the information within DNA into proteins. And rather than physical entities such as clay particles or primitive ribozymes, proteins came to catalyze most of the chemical reactions in the cell.

We know that LUCA existed because all modern cells share a common set of features. One of these features is the ribosome, the molecular machine responsible for building proteins by translating the sequences in nucleic acids into sequences of amino acids. Just like the zircon crystals in Australia that survived Earth's early crust, parts of the ribosome have persisted through billions of years of evolution. They allow us to glimpse a time long before any remaining historical record. Zircon crystals endure because of their physical durability, but ribosomes persist because they have a function so important they have been replicated within every living cell since LUCA. Ribosomes catalyze the polymerization of amino acids to form every protein in every living creature on Earth. Within the active site of every ribosome, we find an RNA molecule doing the work. That RNA, which folds to become a key part of the ribosome, is a living molecular fossil. It exists now with many of the features it had during the age of

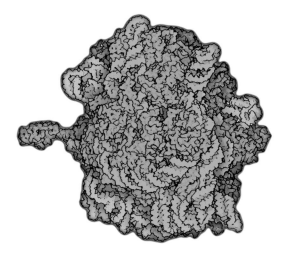

The structure of a bacterial ribosome. The small ribosomal subunit is shown in green and the large subunit is shown in blue.

LUCA, providing us with many clues about our microbial past. The microbes on Earth today that are most like LUCA live in hot liquid habitats.

This is one set of interwoven hypotheses for abiogenesis on Earth: the process of life originating from nonliving precursors. It is a set of hypotheses supported by several intersecting lines of scientific evidence. The hypotheses tell us that life evolved in a freshwater pond near thermal features such as geysers, hot springs, and volcanos. It was a place that may have resembled the landscape of Yellowstone today, like the Grand Prismatic Spring where our modern-day visitor stood with his headlamp under the stars. It was a place that had both the chemistry and the range of temperatures needed to foster life. If not in a terrestrial pond, life might have started near a smoking hydrothermal vent on the ocean floor, although there are complications with that theory. The process of life requires that many prebiotic components of primitive cells first accumulate in sufficiently high concentrations within a viscous solution. How could fatty acids and micelles concentrate and form vesicles if they continuously spewed out of hydrothermal vents and were diluted by the ocean? Too much salt also prevents vesicle formation, and early oceans were twice as salty as they are today.

The reality is that there may have been many pockets on the Hadean Earth where primitive cells took on one or more of the properties of biological organisms. One population of protocells may have evolved, only to be wiped out by an asteroid. Another population may have evolved and persisted in one area for some time, but then was overcome and outcompeted by another, more efficient form of life. Describing life before LUCA, were we able to do so, might require new disciplines, new methods, new categories, and new names to describe all of the exotic protobiologies. Although we remain ignorant of the particulars, we do know that life-forms evolved from microscopic protocells to a microscopic LUCA and then, continuing as microbes, quickly multiplied and colonized every habitable niche.

Life spread across the planet during the Archaean eon and, by 3.7 billion years ago, had already differentiated into many stable species of microbes. The world you would see then might as well be another planet from the one we inhabit today. Lakes and oceans were lined by fields of dome-shaped stromatolite structures. These stromatolites were built by microbes that we traditionally think of as single-celled organisms. However, the stromatolites ranged from the size of a brussels sprout to a head of lettuce or larger. The interior of each stromatolite was not alive; it had a dry texture similar to cement. But the exterior harbored a thriving community of millions of microbes arranged into multi-layered sheets: communities known as microbial mats. Early photosynthetic species formed the foundation of these communities. They converted energy from the Sun into usable chemical forms of energy, just as photosynthetic organisms form the foundation for the biosphere today.

As old stromatolites were slowly buried within the oceans and lakes of the past, they became locked away in the sediment, destined to become fossils. The oldest stromatolite fossils found so far are in 3.7-billion-year-old rocks in Greenland. There are also 3.5-billion-year-old stromatolites in Australia. There are younger stromatolites in several of the national parks of the United States. For example, in Glacier National Park in northern Montana, there are 1.5-billion-year-old stromatolite fossils along the Going-to-the-Sun Road.

And then there are relative "newborn" stromatolites, a mere 200 million years old. These can be found in some remote areas of the United States. You'll have to endure a grueling hike through the desert if you want to see the stromatolite outcrops in Capitol Reef National Park in Utah. These Capitol Reef

A vision of the early Earth. The foreground is a 1902 slide of a thermal feature in Yellowstone National Park later destroyed in a 1959 earthquake. The background is an artist's conception of Pluto.

stromatolites were formed by the activity of microbial mats that lived in a lake oasis during the Jurassic period, surrounded by blowing sand dunes. As the microbial mats lived beneath the Sun, pterodactyls may have been flying in the sky above. Stromatolite fossils like those locked in Capitol Reef rock formations are composed of many laminations of rock that curve upward to form convex shapes. This is where the microbial mat once lived. As the mat was mineralized by the surrounding water and trapped sand particles over time, the microbes moved upward toward the Sun, establishing a new mat where infrared light was more intense.

Stromatolites grew for billions of years with a continuous source of energy from the Sun and without any animals to eat them. It was an era of life on Earth when microbes grew so freely that they were not microscopic at all. Coastal stromatolites grew into reefs. Some of these microbial reefs may have been as large as the Great Barrier Reef off the coast of Australia today. The photosynthetic microbes that built the first stromatolites were replaced by other species

An outcrop of the Altyn formation along the Going-to-the-Sun Road, Glacier National Park.

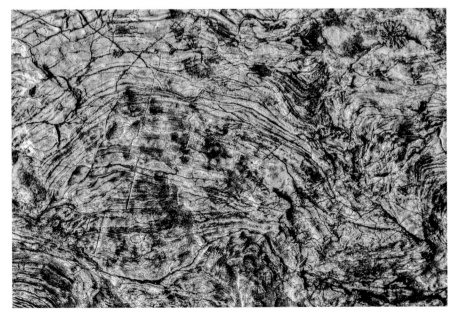

Texture of layered, dome-shaped stromatolite fossils in Glacier National Park.

(Opposite) Glacier National Park, northern Montana, as seen from a satellite.

over time. Early forms of photosynthesis shifted to the type of photosynthesis that we think of now. The ancestors of modern cyanobacteria evolved as early as 3 billion years ago. They used water and carbon dioxide for photosynthesis, producing oxygen as a waste product. These cyanobacteria thrived across the planet. As they released oxygen, they transformed the atmosphere and set the course for future epochs of life. For example, the oxygen produced by cyanobacteria led to the formation of the ozone layer that blocks almost all of the harmful ultraviolet radiation from the Sun. The oxygen produced by ancient cyanobacteria also kick-started a new wave of biodiversity that includes all of the animals that breathe oxygen today.

The rise of cyanobacteria during this period is recorded within Earth's geological history as black and red layers of rock known as banded iron formations. Banded iron formations formed within the sediment at the bottom of ancient oceans. Rocks from the Pilbara region of Western Australia dated at 2.4 billion years old contain one of the largest and most colorful deposits of banded iron formations anywhere. Bands of black iron oxide hematite alternate with red bands of jasper. The black bands formed when oxygen produced by ancient cyanobacteria reacted with dissolved iron, forming iron oxides that precipitated and fell to the ocean floor. Underwater volcanos and other sources provided the accumulations of dissolved iron necessary for this reaction to occur. Nowadays, there is such a steady supply of oxygen in the atmosphere and in the ocean that banded iron formations no longer form. Many varieties of banded iron formations appear all around the world in rocks dating from a period in Earth's history known as the great oxygenation event, when oxygen levels rose rapidly within the atmosphere.

Eventually, microbes became so good at making communities that individual cells became irreversibly dependent on each other, evolving into multicellular organisms. This marked the beginning of the end for the age of stromatolites on Earth. Animals began to evolve over the past billion years, particularly during a period known as the Cambrian explosion, around 500 million years ago, when most modern animal groups appear in the fossil record. And to many of these animals, microbial mats were food. Snails grazed on communities of cyanobacteria and prevented them from forming large structures, such as stromatolites. Even so, there are special places on Earth today where stromatolites continue to grow. These are ecosystems where only microbes can withstand the extremes

(Opposite) Sandstone formations within Capitol Reef National Park, Utah. The Fremont River in the foreground is named after the Fremont Culture, a group of Native Americans who lived in this region.

(Overleaf) A large outcrop of stromatolite fossils is visible in the formation to the right (*fore-ground*). South Cotton Wood, Capitol Reef National Park.

Individual layers within the stromatolite fossils in Capitol Reef National Park show where microbial mats once lived.

of heat, salt, and drought—places where we can peer into the ancient past of life on Earth.

Back in Yellowstone, not far from Grand Prismatic Spring, we find Octopus Spring, where hot waters serve as an ideal habitat for thermophilic microbes that grow into hardened microbial mats. The edges of the thermal streams at Octopus Spring are stromatolite nurseries. Every large stromatolite in these patches was

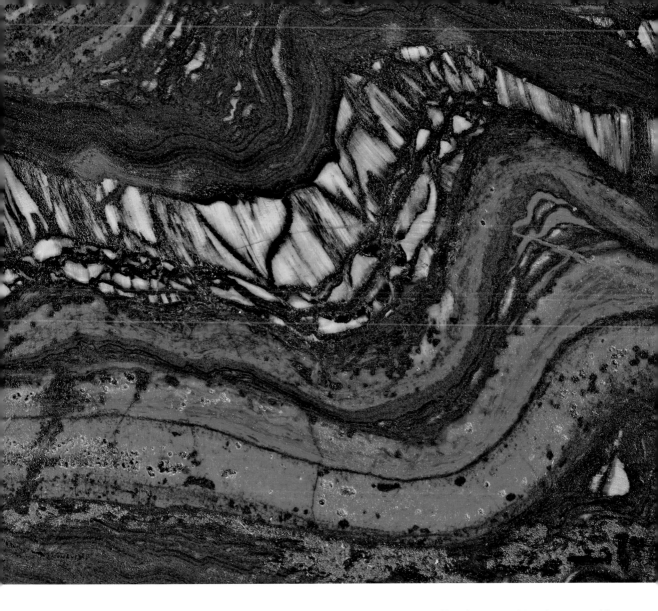

Banded iron formations, 2.4 billion years old, from the Pilbara region of Western Australia.

once one of the barely visible white objects in the thermal stream, smaller than a pearl and glistening in the Sun. The brilliant oranges and yellows are pigments produced by thermophilic cyanobacteria that protect them from ultraviolet radiation. Each stromatolitic structure is shaped over time as the hot, mineral-rich water slowly turns microbial mats into stone. When we look at some of these structures from the side, we can see individual layers of microbial communities.

Octopus Spring, Yellowstone National Park. The spring is over 90°C, driven by surges of water that enter from below every few minutes.

A thermal stream lined with microbial mats flows from Octopus Spring. The hottest area of the stream is the center, where the water is running rapidly from the source. This area is lined with the most thermophilic species, including *Synechococcus*. Other species known to live within the microbial communities in the outflow streams, particularly as the water cools around the edges and farther from the source, are *Aquifex*, *Thermotoga*, and the green filamentous sulfur bacterium *Chloroflexus*.

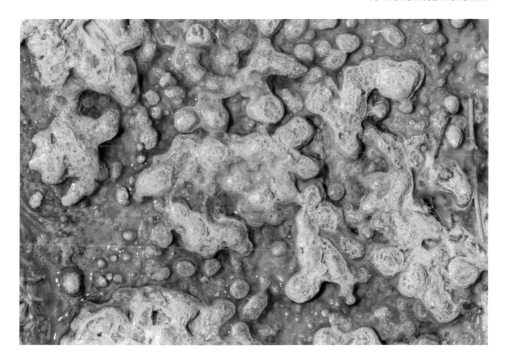

Photosynthetic cyanobacteria are at the top, where they receive the most light. Other species that require less oxygen live underneath them, toward the center layers of the mat, and there is a mineralized white layer at the bottom.

A nursery of growing microbialite structures within the microbial mat.

We are looking at a network of interdependent relationships among microbes living in the mat, with one species producing something that another species needs. Under the microscope, thermophilic cyanobacteria fluoresce on their own from the chlorophyll and other pigments that they produce (a phenomenon known as autofluorescence). By shining ultraviolet light on the microbial mats from Yellowstone, cyanobacteria stand out as bright pink cells. They grow and move within sheaths that appear blue and are surrounded by many other species of various shapes and sizes within the community that do not fluoresce, appearing blue or black against the background.

A similar process is under way south of Wyoming in Utah's Great Salt Lake. At the tip of Antelope Island in the southern half the lake, the salt concentration of the water fluctuates at around 15 percent. Here, microbial mounds are built

Close-up of tiny pearl-shaped microbial structures mineralizing into larger microbialites.

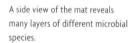

A side view of the mat reveals many layers of different microbial species.

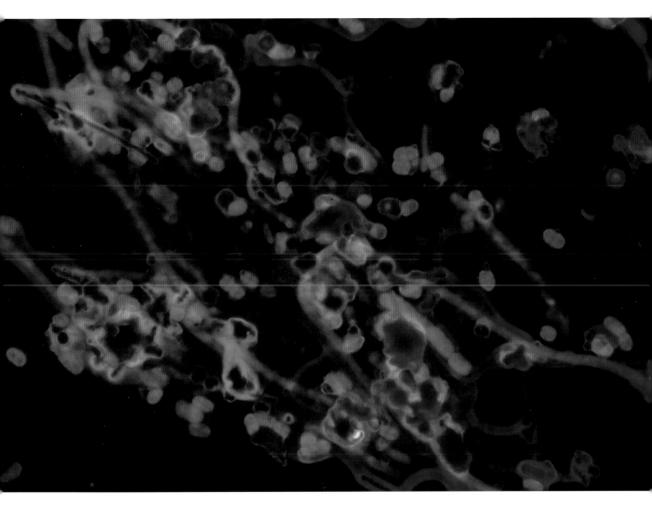

by salt-loving halophilic cyanobacteria, guarded by American avocet birds that make their nests atop the mounds. Like the thermophilic cyanobacteria in Yellowstone, looking at one of these salty, microbial mounds from the side exposes layers of organisms producing green, brown, and yellow pigments. The microbes that produce these pigments are among the many different bacterial, archaeal, and eukaryotic species that live within the mats. The activities of these microbes cause limestone and other minerals to precipitate out of the brine, forming reef-like microbialite structures over thousands of years.

A community of thermophilic microbes. The cyanobacteria fluoresce pink under ultraviolet light.

In some corners of the southern part of Great Salt Lake, water evaporates along the shore and raises the salt concentration even higher, creating the perfect conditions for a community of halophilic archaea. These haloarchaea live among other microbes in a muddy substrate, cracked by strong sunlight. It is a dense community, owing its deep red color to carotenoid pigments produced by the haloarchaea: the same pigments that give the red color to the northern half of the lake, which also has a high concentration of salt (above 25 percent). Looking closer, we see the texture of the red microbial mat, crossed by a streak of yellow algae. Zooming even closer using one type of light microscopy called bright field, we can see the natural colors of each organism, including the makers of the carotenoid pigments themselves: pink, rod-shaped haloarchaea from the genus *Halobacterium*. With another type of light microscopy, phase contrast, the precise shapes and sizes of the many different cells are emphasized. And using electron microscopy to see the smallest size scales within the mud, we see dense biofilm clusters of smaller bacterial and archaeal cells. The larger cells embedded within the biofilm are diatoms—one of the most common photosynthetic eukaryotes on Earth, with a glassy cell wall made of silicon dioxide. Some of the first species on Earth probably colonized the many hypersaline lakes similar to Great Salt Lake that likely covered portions of the early planet.

The chemistry that laid the foundation for life on Earth developed in a stepwise fashion across many different locations. Let's assume the fatty acids that became cell membranes of the first cells on Earth were manufactured on a geyser wall. Then those fatty acids were blasted out of the geyser, propelled by heat within underground magma chambers, and carried off by a cool wind. Perhaps the nucleotides that made the RNA in those protocells came from cyanide at the bottom of a lake. Where did the energy come from in each chemical reaction along the way to produce that cyanide? Was it ultraviolet light from the Sun? Was it an asteroid entering the atmosphere? A bolt of lightning? Was the cyanide circulated through streams filled with rain water? Did the water vapor that formed those raindrops come from the ocean? Were some chemical precursors picked up in a cloud? All of the molecules critical for the success of the earliest cells have a history spanning time and space.

In every location, at every step in the origin of life, the physical environment was a key contributor. The environment might have been the catalyst of the chemical reactions themselves. It might have been a part of the cell's replication

(Opposite) The upper end of Great Salt Lake, Utah, is colored red due to the natural pigments of halophilic microbes.

(Overleaf) Microbialites grow in the salty waters off Antelope Island.

Green cyanobacteria form layers within the Great Salt Lake microbialites.

cycle, providing the mechanical force or temperature changes for cells or molecules to separate. If life evolved in a geothermal pond comparable to those in Yellowstone today, that pond's very physical properties and volume played a role by determining the concentration of all of life's ingredients. If just one of these fleeting events had not occurred, the cell at the end of the storyline would not have succeeded.

So when we ask where life first evolved, what we really mean to ask is a whole series of questions. Where were the primitive cells of life when they took that very last step? Where was the protocell with a stable membrane capable of dividing, with stable RNA as genetic material? In what habitat did that protocell arise? Where was the protocell when it began undergoing Darwinian evolution? Where did it develop a protein-based metabolism—when it began to sense and change the environment around it? We may never know. The best chance to understand life at its infancy is to evoke the concept of conservation of biology on Earth similar to the conservation of energy in the universe. The same precursor molecules of early life must be manufactured someplace on Earth today. We can

Water evaporates in a corner along a beach on Antelope Island, increasing the salt concentration and promoting growth of halophilic archaea.

go to geysers and hot springs and analyze how the molecules of life are forming, how they are distributed, where they are gathering, and how they are interacting with other molecules. We can mimic this chemistry and try to evolve life in the laboratory. If we ask where life evolved, we must be prepared for a big answer. Life evolved here on Earth. Life evolved from matter once collected in a solar nebula that collapsed long ago to form our Solar System.

A biofilm of red halophilic microbes.

Texture of the mud biofilm with a streak of yellow algae.

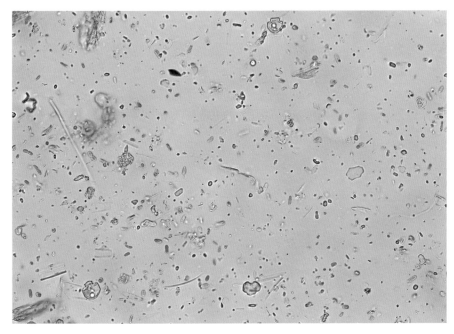

The natural pigments of Great Salt Lake microbes, including pink haloarchaea, seen in Great Salt Lake mud diluted in a salt solution.

(Overleaf) Green diatoms and other halophilic microbes from Great Salt Lake mud.

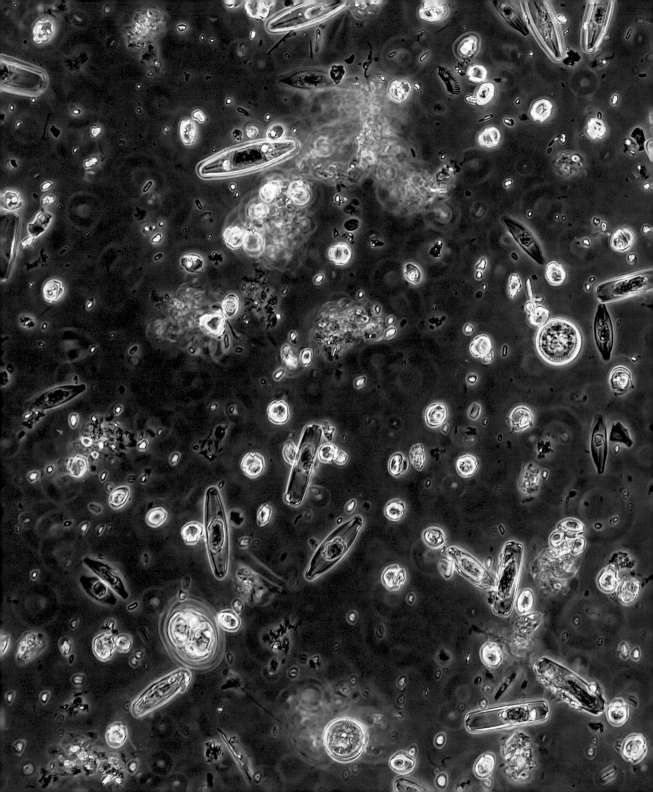

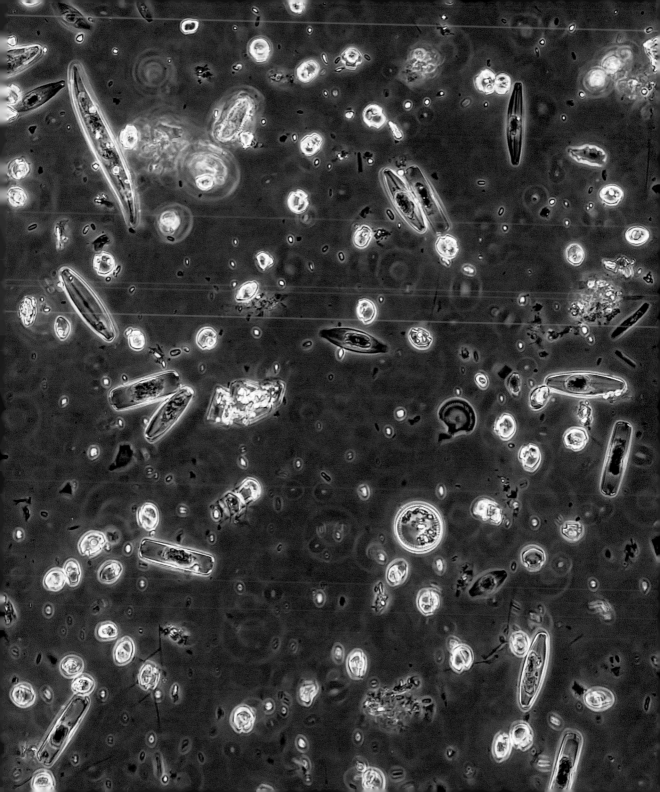

Here, powered by the Sun, is the biosphere of Earth. Here are all of the microbes and all of the plants and animals that arose much later under a protective ozone layer produced by microbes. Here, where this story began in Yellowstone National Park, is a pack of gray wolves running in the mountains above Lamar Valley at sunset. As darkness falls, the wolves begin to stalk a group of American bison. The bison stand at the top of a ridge, half-asleep and completely unaware of the wolves, who have encircled a newborn calf. With a signal from the alpha male, the wolf pack bursts into action. Racing up the hill, the wolves' muscle cells are powered by mitochondria—the bacterial endosymbionts that joined with another microbe long ago to form the eukaryotic cell. The wolves tear the calf from the herd, but the calf fights back. With every breath in the bison calf's

Lamar Valley at sunset, Yellowstone National Park.

(Opposite) Great Salt Lake mud biofilm as seen with electron microscopy. The large, glass-shelled cells are diatoms. Many smaller species of bacteria and archaea live within the surrounding biofilm.

An American bison rests in Lamar Valley.

lungs and every beat of her heart there are oxygen molecules produced by photosynthetic microbes in the ocean. Finally the calf is free to return to the safety of the herd. She lives on in the valley.

Pause to marvel at the interconnectivity of life. The bison and every other majestic animal evolved from microbial ancestors. All animal actions today are driven by microbes. Animals are driven by the microbes that produce oxygen in our atmosphere, by soil microbes that give life to the grass that gives life to the bison, and by the energy first taken into that grass by another microbial endosymbiont, the chloroplasts of plant cells. Bison depend on trillions of bacteria living in their gut to digest the grass. And those microbes in the gut produce neurotransmitters capable of changing mammalian behavior. Maybe it was also the microbes that gave the calf the strength to survive.

3

Under Celia Thaxter's Garden

TEN KILOMETERS OFF THE COAST OF MAINE, the poet Celia Thaxter stepped into her flower garden on Appledore Island holding the hand of her young grandson Eliot. Ever since her first small garden, planted when she was Eliot's age, Celia was delighted by flowers and the little animals that visited them. All were guests at her family's storied 1800s hotel, Appledore House, located on the Isles of Shoals. "Welcome, a thousand times welcome, ye dear and delicate neighbors—Bird and bee and butterfly, and humming-bird fairy fine!"

Outside the house, through the open gate of the garden, the Impressionist painter Childe Hassam sat before an easel with his paintbrush brought to a fine wet tip. With a dab into the red, Childe painted the plot of poppies. With yellow he added honeysuckle, and with shades of green he painted all the stems. He painted Celia and little Eliot, too. There, blended at the center of one colorful composition, Celia, Eliot, and Appledore House appear as if they are all a part of the living garden. Childe took the painting into the parlor of Appledore House and hung it to dry, forever capturing this special place.

Throughout her years as hostess, Celia's parlor was a bustling center of creativity. Harriet Beecher Stowe, Ralph Waldo Emerson, Henry David Thoreau, Nathaniel Hawthorne, and many other writers, painters, and musicians of the day convened there to enjoy song and drink, looking out on the garden and breathing in its fragrance. The garden was the heart of the property. It was the source of Celia's inspiration and the focal point around which all of these literary and artistic luminaries came and went.

But the garden should not have thrived where it was. Appledore Island is small and windy, with rocky ledges that barely rise above the Atlantic Ocean. The ground on the island is not fertile. Back on the mainland the most common biome is deciduous forest. This same community of plants and animals found in New England covers the eastern half of North America and most of Europe. In deciduous forests, rich soils develop over millennia as leaves fall from trees and decompose into the earth. On Appledore Island, however, there are no forests.

Celia and Eliot Thaxter in *The Garden in Its Glory* by Childe Hassam, 1892.

Appledore House on the Isles of Shoals as it was in the 1890s.

In fact, an early English explorer on the Isles of Shoals wrote, "Upon these islands I neither could see one good timber-tree nor so much a good ground as to make a garden." It's no wonder he said this, because the biome on the Isles of Shoals is different from the deciduous forest of the mainland. It consists of coastal grass and shrubs. Only the hardiest plants live there, and they do not produce a naturally fertile soil.

The garden, a living work of art, existed on Appledore Island because of Celia's hard work. Every autumn, she began tending her garden for the following spring. The "first necessity," she wrote, "is the preparation of the soil." For this job, Celia felt lucky. She knew where there were heaps of barn manure on the island, which had been left for so long they had become a "fine, odorless, velvet-brown earth, rich in all needful sustenance for almost all plants." Over the years she learned just how much barn manure to sprinkle over each species of plant.

The garden was Celia's own small ecosystem. The soil there was just as rich as it was under the deciduous forest on the mainland. Barn manure replaced fallen leaves, rendering the earth fertile. But this is not the end of the story. Decomposition of organic materials in soil is an active process. Everywhere that Childe Hassam painted a colorful garden, another ecosystem thrived underground.

Like deciduous forest, coastal scrubland, and other major biome types, each pinch of garden soil is the bedrock for microbiomes. Instead of spreading across entire countries and continents, microbiomes spread between microscopic particles of dirt. Instead of being controlled by the climate, by trade winds and weather patterns, they respond to a single drop of water, to one fallen leaf or a sprinkle of manure. These microbiomes are hidden ecosystems.

Soil microbiomes contain large numbers of individual species. There are many more species in the soil microbiome than there are species of animals in a tropical rainforest, including all of the insects, reptiles and amphibians, jaguars, and poisonous frogs. The biologist E. O. Wilson once counted forty-three species of ant on a single bush in the Amazon rainforest: more species, he famously guessed, than "the entire ant fauna of the British Isles." We can take this a step further with microbes. Within a single gram of soil surrounding the roots of Wilson's bush, there are tens of thousands of different microbial species: more microbial species than the number of ant species on Earth.

Ten billion microbial cells live in a gram of garden soil. When we smell dirt, what we are detecting is the smell of microbes. The characteristic "earthy aroma" of a garden or a forest after a rain shower is caused mostly by a single molecule produced by many soil microbes. The molecule is geosmin, derived from the Greek roots *geo* (earth) and *osmo* (smell). But geosmin production is just one of innumerable chemical reactions catalyzed by the invisible soil inhabitants. When a new source of organic material, for example, a falling leaf, is added to the soil, microbes respond by releasing extracellular enzymes into the environment. Exoenzymes break down complex organic compounds into simpler molecules that they can utilize as nutrients and energy sources. In the process, these simpler compounds become available for other organisms in the soil. Without microbes to break down manure and leaves on the surface of the soil, organic materials would decay very slowly, and the cycling of carbon through the environment would be stuck in gridlock. Most soil microbes are found either near the surface where fresh organic material is deposited, or in the area surrounding plant roots—known as the rhizosphere—where they eat nutrients produced by the plant and break down dead plant cells.

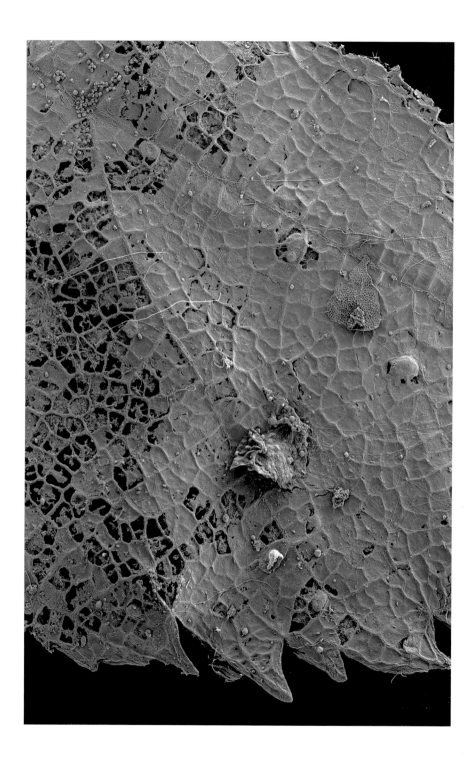

A leaf on the forest floor
is slowly decomposed by
microbes.

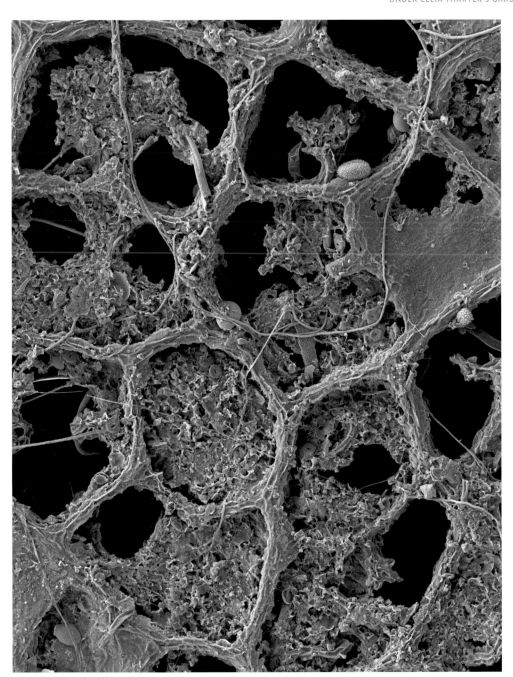

Fungal hyphae, bacteria, and spores persist among the remaining veins of the leaf.

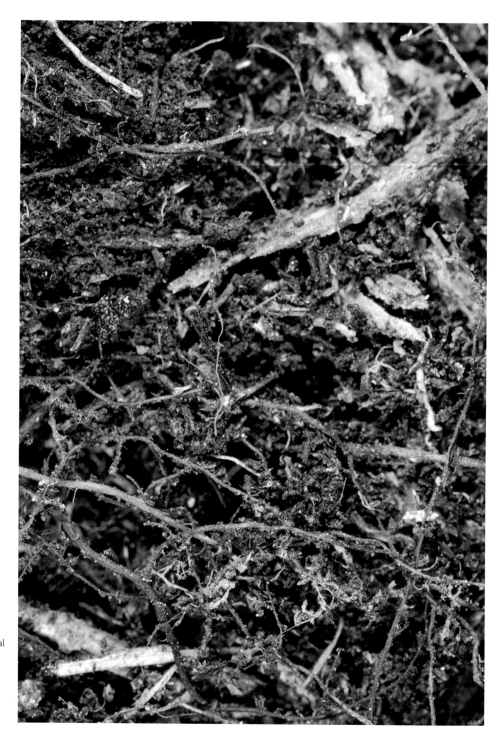

Soil is rich with microbial diversity.

What are these abundant microbes that co-exist in the neighborhoods of fertile soils? The most abundant are the bacteria, and among these stand out the geosmin-producing *Actinomycetes*. Every gram of soil also contains many fungi in the form of meters of microscopic threads called hyphae. Each thread is not much wider than a typical bacterial cell, but hyphae can bundle into thicker strands called mycelia that are strong enough to push through soil particles. Fungal mycelia collect water from the soil and secrete exoenzymes that break down nearby organic material. These networks, though barely visible to the naked eye, are among the largest organisms in the world. In Oregon, one contiguous

White fungal hyphae form networks within the soil.

mycelial network formed by the honey fungus *Armillaria ostoyae* covers ten square kilometers. This "humongous fungus" extends up into tree trunks as it expands its forest footprint, living beneath the bark and emerging as mushrooms every autumn.

Through these ancient mycelial networks, fungi form symbiotic relationships with plant roots in the soil. These symbioses are mediated by a sticky substance that roots secrete into the rhizosphere to attract fungal mycelia. It starts on a sunny day, meters above ground, as photosynthesis inside the cells of the leaf capture carbon dioxide from the atmosphere, producing simple carbohydrates, sugars like glucose. The sugars travel down the stem and into the roots, where they are released into the rhizosphere and taken up by mycelia. These networks of fungal mycelia in tight association with plant roots are called mycorrhiza. It's a mutually beneficial symbiosis: the fungus receives a consistent supply of sugars from the plant, and the plant gains better access to water, minerals, and phosphorus delivered by the mycelial network.

Mycorrhiza are subterranean superhighways of exchange that link the metabolisms of nearly every organism within the soil. In some mycorrhizal connections, mycelia contact the outermost cells of a plant root. But most of the time, fungal hyphae actually penetrate the plant tissue, becoming a physical part of the root. These networks are so large and connect so many different categories of life that it is difficult to tease apart where one organism ends and another begins. Studies in forests have shown that individual plants exchange carbon and other nutrients with each other through fungal networks. This occurs between the largest old-growth trees and the smallest saplings, even across different plant species. Mycorrhiza fossils have been found in rocks over 400 million years old. It is easy to imagine the entire forest ecosystem as one resilient biological entity.

Along with the mycorrhizal connections, some of the plants in Celia's garden formed a special kind of symbiotic relationship with nitrogen-fixing bacteria, bacteria able to convert nitrogen into ammonia. The sweet peas in her garden are legumes, and legumes have structures on their roots called root nodules. Each root nodule is filled with a colony of millions of bacteria known as rhizobia, meaning "life in the roots." When rhizobia are associated with legumes, they capture nitrogen gas from the atmosphere and convert it into ammonia. Many bacteria are able to carry out this process of nitrogen fixation while living independently, but those that form this intimate symbiosis with legumes provide

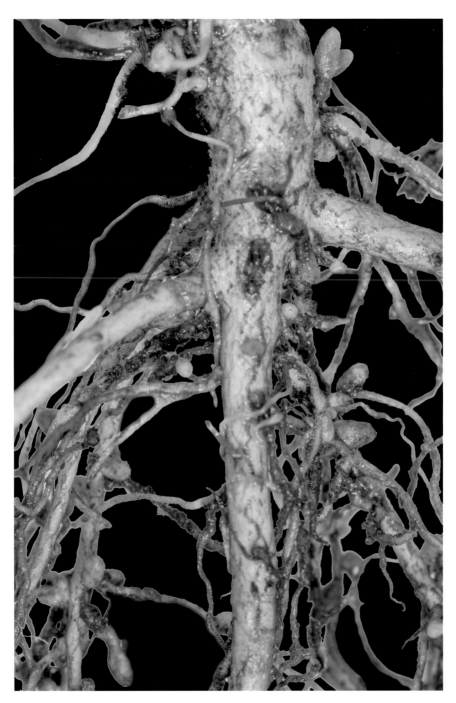

The root nodules of a legume plant, a clover from the genus *Trifolium,* contain symbiotic bacteria.

The pink hue of root nodules
comes from the protein
leghemoglobin.

the benefits of nitrogen fixation directly to the plant. Root nodules appear pink to the human eye because they contain leghemoglobin, a protein similar to the hemoglobin in our blood but which serves a very different function. While our hemoglobin transports oxygen to our organs, leghemoglobin keeps oxygen away from the enzyme complex that converts nitrogen gas into ammonia. Oxygen would inactivate the enzyme complex in the rhizobia.

Nitrogen is the fifth most common element in the universe and is one of the essential elements of life. On Earth, most nitrogen is held within molecular nitrogen gas, which makes up 78 percent of the atmosphere. However, nitrogen gas is chemically inert; it's in a form that is unavailable for use by most organisms. For this reason, many ecosystems are limited in nitrogen more than any other nutrient. These ecosystems depend on nitrogen-fixing bacteria to produce ammonia as a usable form of nitrogen. Nitrogen is assimilated into amino acids for making proteins, and nucleotides for making nucleic acids and many other biological molecules. Plants with root nodules grow even when there is little available ammonia in the surrounding bulk soil. As these plants die, the nitrogen in their cells is decomposed by other bacteria and fungi, releasing usable forms of nitrogen within the soil as a natural fertilizer for other plant species. Meanwhile, other bacteria in the soil continuously denitrify biological compounds and release fresh nitrogen gas into the atmosphere, completing a balanced nitrogen cycle.

The bacteria and fungi living near plant roots attract the predators and grazers of the microbial world. Bacteria are easy prey for microscopic animals such as nematodes, tiny worms that eat smaller microbes. And with no shortage of bacteria to eat, nematodes are one of the most successful animals on the planet. Nematodes are found in every part of Earth's lithosphere, from topsoil to environments kilometers below the surface. They range from microscopic to large enough to see unaided, a few millimeters long.

Bacterial cells contain more nitrogen than nematodes need to survive. When nematodes and other predators eat bacteria, they release excess nitrogen in the form of ammonia. Like the ammonia produced by nitrogen-fixing bacteria, this is exactly the form of nitrogen that plants require to grow. So every time a nematode eats a bacterium, it is supporting not only itself but the lives of all the plants living within that patch of soil. Multiplied by the number of predators and the thousands of bacteria each predator eats per day, this process produces much of the bioavailable nitrogen in the soil.

This little pill bug, a terrestrial crustacean more closely related to crabs than to insects, feeds on decaying plant material within the soil food web.

The soil microbiome is a deep, dark world, red in tooth and claw. Scavengers course through the topsoil, consuming fungi and bacteria as they shred dead plant material. These are some of the creatures of childhood nightmares: millipedes, pill bugs, earwigs, and earthworms. There are no laws governing who eats whom and when. Fungi can turn the tide on nematodes that eat fungi, becoming predators themselves. These predatory fungi form specialized nets and loops that trap nematodes.

The sum of these complex ecological connections between organisms allows many thousands of species to live in close proximity. There are hundreds of different habitats for microbes within a handful of garden soil that looks like nothing more than a simple pile of dirt. There are many different ways to live in each of these microhabitats: different ways of adapting to survive on a grain of sand, on a root, or on a speck of manure. We have seen decomposers, scavengers, grazers, predators, and prey. Together, they form the soil food web. The soil food web encompasses all of the underground microbes, all of the plant roots, the earthworms, and the creepy, crawly insects. This web extends above ground to the bees that gather flower nectar and cross-pollinate plants, to the birds that eat earthworms, and to people. After all, humans grow and survive by eating plants or animals that eat plants. Soil is the largest living reservoir of biodiversity on Earth.

Celia Thaxter never knew of the microbiomes in her garden. What would she have thought if she had known that all of her flowers were connected through an underground network of fungal mycelia? Would she have written poems about nematodes and bacteria if she had seen the microbial world?

Celia's exploration of nature lived on through the work of one her sons, Roland Thaxter. Roland grew up running through the maze of his mother's summer flower garden and went on to become a talented student of the natural sciences. He studied at Harvard in nearby Cambridge, Massachusetts, where he became a professor of biology and botany. Roland spent the summer recesses with his children in Kittery Point, Maine, where his father, Levi, owned property. Roland continued his scientific research in Kittery Point and spent time with his children and mother right across the water on Appledore Island.

Roland's affinity toward nature motivated him as a scientist and an artist. Instead of growing flowers as his mother did, he sought out natural forests. Roland used his artistic talent to make exceptional drawings of the life he discovered in the woods. His research focused on fungi; he wanted to find and understand the small and hidden life of the forest—living forms that were barely visible to the naked eye.

Let's take a walk through the very same woods in Maine where Roland Thaxter worked every summer. The trail leaves from his cottage on the coastal peninsula of Cutts Island, snakes through a dense, sunlit forest, passes through a dark swamp, and opens into a salt marsh leading to tidal pools on the beach. We'll see many of the same things that Roland saw, but we'll use modern microscopes to zoom in even farther. We'll begin with macroscopic landscapes and move on to light microscopy, then to electron microscopy. We'll make use of the current understanding of microbial diversity for a glimpse of each microbial ecosystem and look closely at a few of the individual microbes along the way, as if we could shrink down and enter each microbiome we encounter.

Heading down a trail into a sunlit forest in Kittery Point, Maine.

Of all the objects in the forest, the most noticeable are the leaves that hang from the trees. The surface of each leaf seems smooth to the touch, but every leaf has thousands of bacteria and fungi on its surface. Some of these might have been blown there, carried by the wind. Even the air of the forest is a microbiome. We can see that there are microbes on the leaves without using a microscope. To do this, we take a leaf and press it against a nutrient-dense agar surface inside a petri dish. Then, over a few days, the microbial cells that were transferred from the leaf will grow, from one cell to two cells, two to four, four to eight, and so on, exponentially, until there is visible growth in all of the places where microbial cells were transferred from the leaf onto the agar surface. The colonies that grow from a fern leaf and an oak leaf form spatial patterns reflecting where microbial species were living on the original leaves as they were in the forest. On the oak leaf print, there are green *Trichoderma* and pink *Fusarium* fungal colonies.

Many of the same species from the soil will be found on decaying logs along the forest trail. At first we are drawn to a blue-green lichen colony growing on the surface of a log: a composite organism made of a fungal species (the mycobiont) and a photosynthetic cyanobacterial species or an algal species (the photobiont), or sometimes both cyanobacteria and algae. Slicing through a lichen and examining the cross section through light microscopy, we see fungal hyphae in blue where they make contact with the photobiont, in this case algal cells stained red. Moving one level closer with scanning electron microscopy, we see the larger, round algal cells and the filamentous fungal hyphae deep within the photobiotic layer of the lichen. Here we can appreciate how symbiosis works—the moment of contact between the mycobiont and the photobiont as the hyphae wrap around the algae, taking up nutrients brought into the system by the algae through photosynthesis. Underneath the surface of the log where the lichen lives, soft decaying wood is being slowly decomposed into the soil by bacteria and fungi. Every so often, fungal mycelia from within the wood develop into fruiting bodies on the surface of the logs, resulting in the mushroom caps so familiar to us.

As we go along our walk, we soon encounter a swamp. Looking deeper into the swamp, we see the plants that thrive in these wet conditions. There are mosses and ferns submerged on water-logged branches. Green moss branches are like smaller forests within the forest: a hall of mosses. Each leaf is covered by thick water droplets. Even the leaves that look dry are covered with a microscopic

Colonies formed by diverse microbes living on a fern leaf.

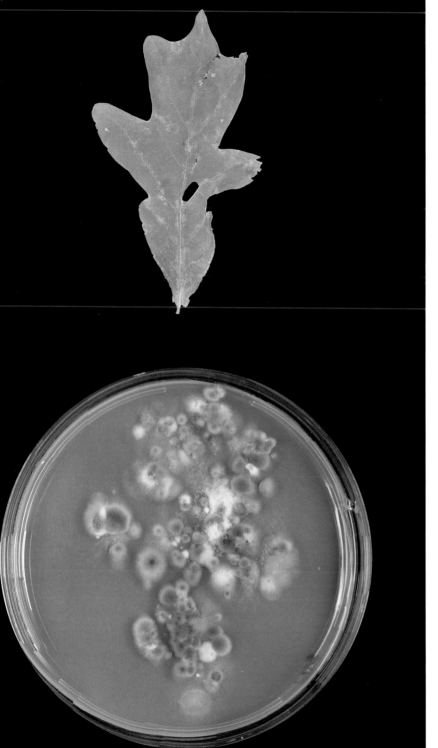

Microbial biodiversity on an
oak leaf.

The forest floor is a thick layer of leaf litter.

A colony of lichen grows on a decaying log.

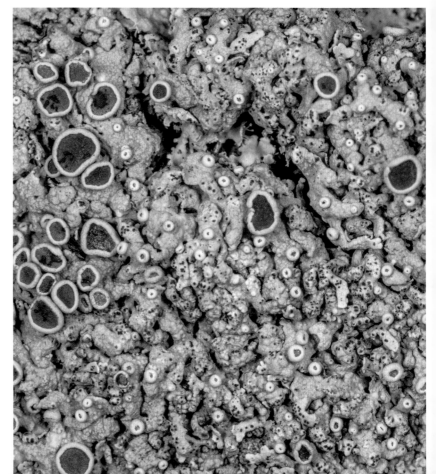

(Opposite) A cross section of lichen with algae cells, stained red, and blue fungal hyphae.

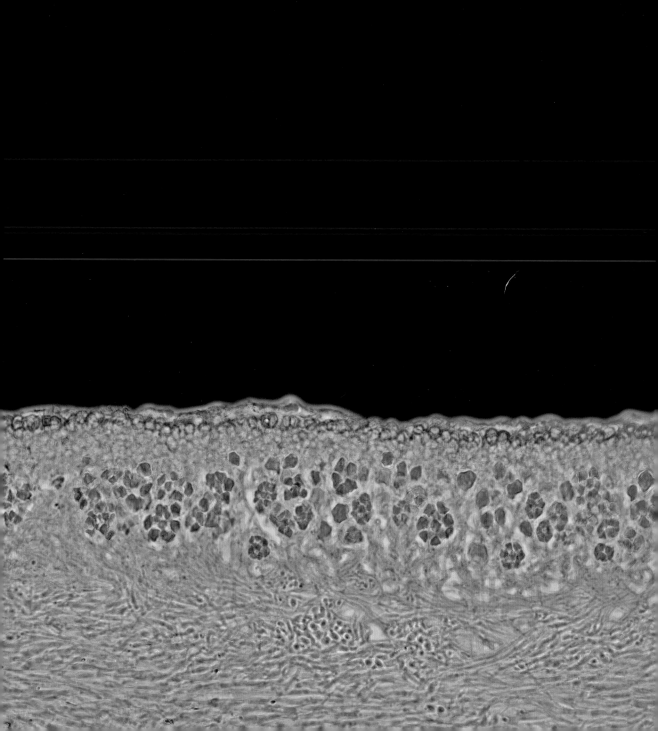

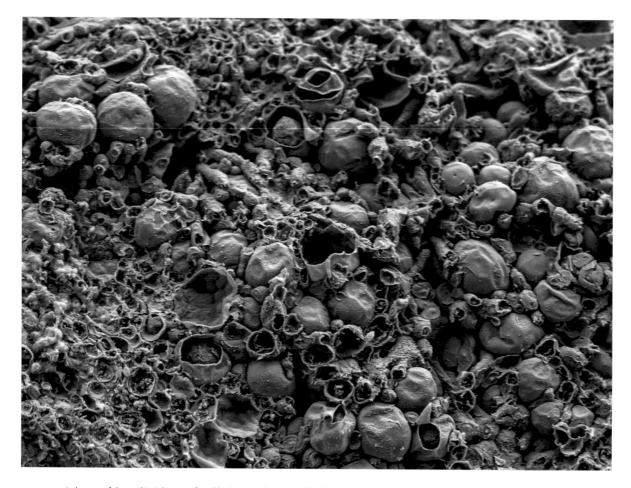

A close-up of the symbiosis between fungal hyphae and algal cells inside of a lichen.

(Opposite) These mushroom caps from the genus *Mycena* are the fruiting bodies of fungal mycelia within the log.

film of water that supports abundant microbial life adhered to the plant tissue. Between tiny grains of quartz sand trapped on the leaves, there are microscopic animals and cells from all three domains of life, scattered about in various shapes and sizes.

The largest of these organisms are one of two types of microscopic animals, the rotifers or the tardigrades. These share a lot in common with nematodes and

The swamp ecosystem.

Every drop of water within the moss is a microbial universe.

(Opposite) Scanning electron microscopy reveals the many microbes within the moss.

Microbial life adheres to moss leaves.

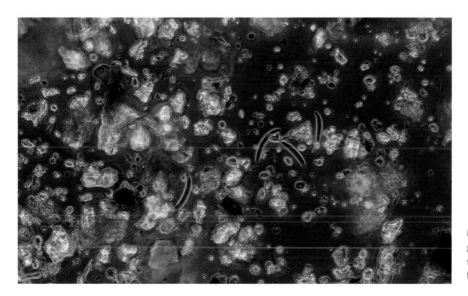

Moss microbes scattered among grains of sand. The crescent-shaped microbes are spores of the fungal genus *Fusarium*.

with each other. First, they are around the same size. If we place tardigrades or rotifers along the lines of a ruler, we would have to line up around one hundred of them to cross one centimeter. They both have pet names: tardigrades are "water bears" and rotifers are "wheel-bearers." Tardigrades are the bears of the microbial world, despite their eight legs, because they walk along moss leaves with a lumbering, bearlike gait. Rotifers are called wheel-bearers because their large mouth lined with beating cilia, called a corona, appears to spin around in circles. There are many different species of tardigrades and many species of rotifers. Herbivorous tardigrades use their tubular mouth, which contains sharp stylet structures, to pierce algal and plant cells. Other tardigrade species are bacterivores. The large oval structure seen behind the mouth is called the pharynx; it creates suction for the mouth. Tardigrades and rotifers develop into adult bodies with a defined number of cells. The anatomy of their transparent bodies is open for observation by anyone with a light microscope. In addition to the structures mentioned, they have simple brains, muscles, digestive systems, and different reproductive organs for males and females. These micro-animals living on a moss leaf could see their entire world evaporate with one gentle breeze. But it is no matter to them because they have evolved a very neat trick known as cryptobiosis. When a local drought occurs, they lose almost all of the water molecules in their body, adopting a dormant, dehydrated state. They can survive in

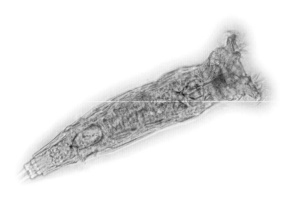

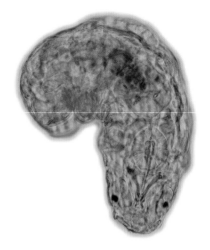

A rotifer from the genus *Philodina*, its corona mouth lined with beating cilia (imaged live).

A tardigrade, sp. *Hypsibius dujardini*, seen with eye spots, stylets, pharynx, and midgut full of algae (imaged live).

The tardigrade *Ramazzottius varieornatus* reanimates from a smaller, dehydrated state of cryptobiosis. The adult individual shown is approximately 300 micrometers long, and the dehydrated tardigrade is approximately 100 micrometers long. Tardigrades can survive many physical extremes in this form, whether temperatures from −200°C to almost 100°C, high doses of ionizing radiation, high pressure, or even exposure to space in low Earth orbit.

suspended animation for years. Then one day, a single drop of water may rehydrate and reanimate them, and they are once again swimming and hunting down bacteria in no time.

We now leave the swamp behind and continue walking through the forest until we reach a salt marsh at the end of the trail. There is a large clearing with a river that flows into the ocean on our right. We walk out onto the marshland to a dry ridge that cuts across to the beach. On either side of the trail there are

The salt marsh at the end of the trail.

Spongy green patches within the salt marsh.

strange pockets of marsh where grass does not grow. The areas feel spongy and are not strong enough to stand on.

When we remove a slice of the marsh, we find several colorful layers beneath the surface. Each one of these layers is a few millimeters thick and is made of a different composition of microbes, forming a microbial mat. A thin, golden-green layer of algae coats the surface at the very top. Beneath this is a blue-green layer of photosynthetic cyanobacteria. Beneath the cyanobacteria is a layer made up of purple sulfur bacteria, though the band appears pink because of carotenoid pigments. Like the cyanobacteria, the purple sulfur bacteria are photosynthesizers. They live below the cyanobacteria because they require only a little oxygen or no oxygen at all. When yellow light is shined on microbial mats, the pigments of these photosynthetic bacteria in the top layers fluoresce red. In the electron micrograph we can see a nematode meandering through the matrix of the mat. Underneath these top layers there is a gray zone of methane-producing

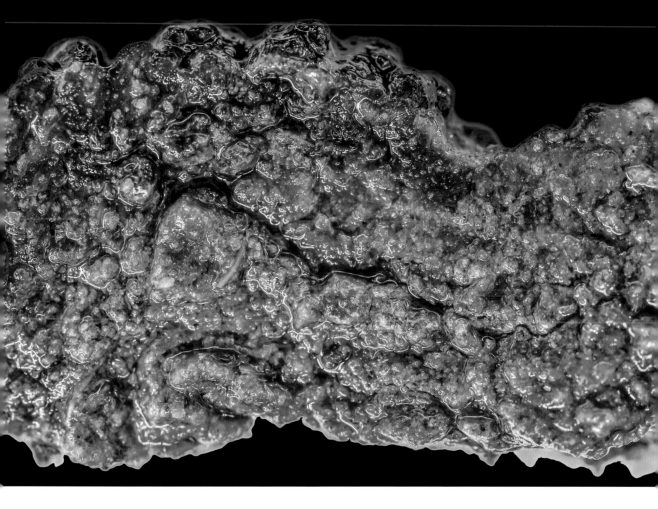

archaea. These methanogens are also anaerobic. Similar methanogenic archaea live in human intestines.

It is time now to move clear of the marsh and toward the ocean. Even before arriving on the beach, the rich smell of low tide drifts through the air, itself a cocktail of hydrogen sulfide and dimethyl sulfide gases produced by beach bacteria. Tidal pools are filled with seaweed and smooth gray and white rocks. Like the freshwater swamp back in the woods, these tidal pools are little universes of aquatic life. Looking at tidal pool water with a microscope, we find a zoo of microscopic animals and algae, including a tiny orange crustacean, a copepod. If we zoom with higher magnification into the black areas surrounding the micro-animals and shine yellow light, the tidal water microbiome glows with an array

A slice through the spongy marsh microbial mat at Kittery Point, Maine. Similar microbial mats are found in Great Sippewissett Marsh in Woods Hole, Massachusetts.

Red autofluorescent cyanobacteria within salt marsh microbial mats.

of bright autofluorescent pigments. Red autofluorescence comes from the chlorophyll in photosynthetic species.

There is yet another microbial community within the minute crevices on the surface of each tidal pool rock, rich with bacteria and diatoms. The glass shell of each of the larger diatom cells in this microscopic aquascape is made of two halves called valves. The elaborate patterns on the valves as seen with electron microscopy are perforations in the shell through which the cell conducts all of its environmental sensing and nutrient exchange, since the glass is impervious to water. We also notice a new organism that lives on the tidal pool rock, a paramecium. This paramecium is not as large or as complex as a nematode, tardigrade, or rotifer. Paramecia are not microscopic animals even though they are eukaryotes. Rather than having a few thousand cells, their bodies are made of one large

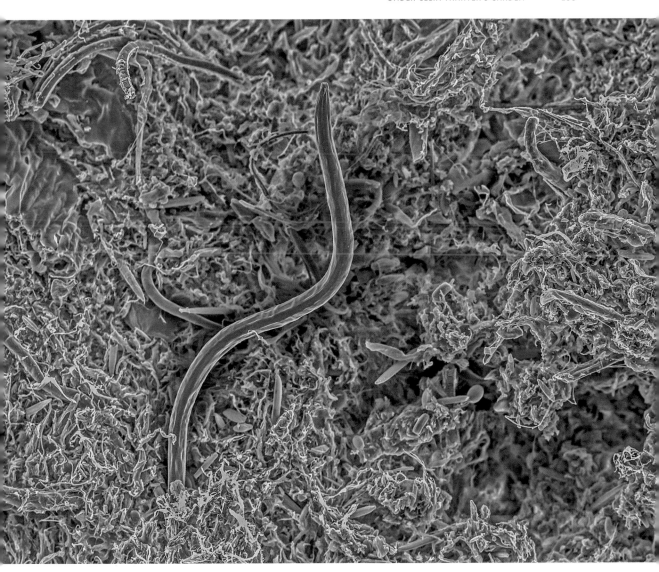

A nematode meanders through the matrix of the microbial mat.

Tidal pools on
the beach at
Kittery Point,
Maine.

Smooth rocks
and algae line
each shallow
tidal pool.

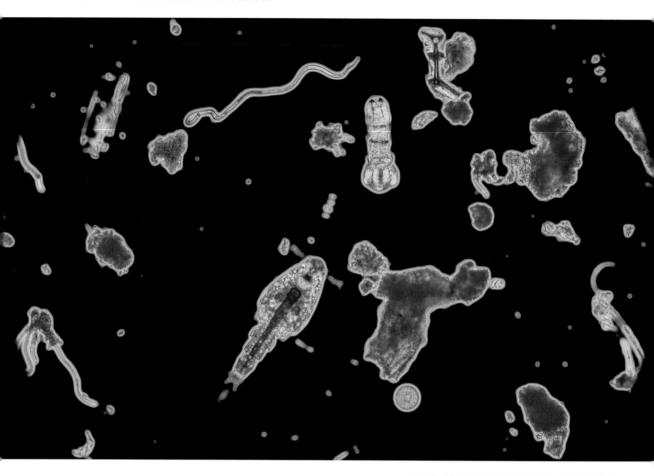

There is a zoo of micro-animals and algae in every tidal pool.

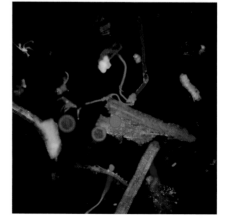

The autofluorescent pigments of tidal pool microbes.

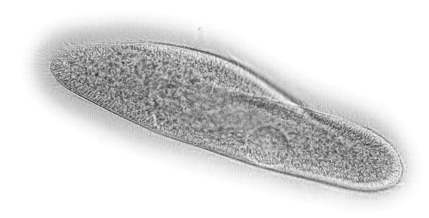

A paramecium, *Paramecium caudatum,* propelled by beating cilia, seen with its oral groove and nucleus (imaged live).

cell. Instead of different tissues, paramecia have organelles inside their cells that carry out important functions. These types of eukaryotic organisms are called protists.

Paramecia belong to a group of protists known as ciliates. They are called ciliates because they are covered by cellular extensions called cilia that beat rapidly and enable rapid forward motion. A structure running down the center of the paramecium called the oral groove acts like a mouth. Paramecia use the cilia along this groove like conveyor belts to bring bacterial prey into their mouths. Once bacterial cells are inside of the cell, they are packaged into organelles called vacuoles that serve as the cell's digestive system. Food is broken down inside of vacuoles, and then the digested contents are released into the cytoplasm. The large, circular organelle seen inside is the macronucleus.

When we arrive at home later that day, there is peace and quiet away from the tidal pools on the beach. Even so, the wilderness adventure continues in our houses and on our bodies. A house is a built environment for microbes and thus

The microbial aquascape in the crevices of a submerged tidal pool rock.

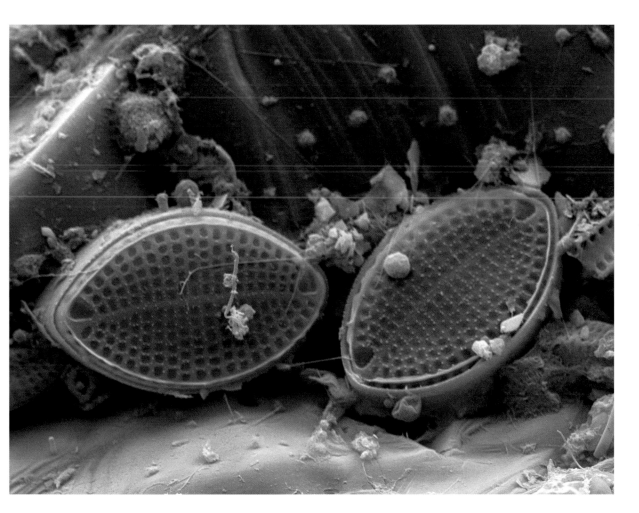

The perforated glass outer shell of diatom cells.

has its own microbiomes. There are microbes on most surfaces no matter where we are, almost all of which are not harmful in any way. In fact, the microbes on our home surfaces are very similar to those living on our skin. One way of seeing some of the microbes on the skin is by pressing our fingers on an agar growth medium, just as we did for the leaves from the forest. Over a few days of incubation, orange, yellow, and white colonies develop from the exact location of each finger's imprint. Our own microbial cloud of mostly harmless bacteria follows wherever we go and colonizes the surfaces of our surroundings.

We've entered only a few of the millions of microscale environments on Earth. The closer we look, the more biological diversity there is to see. Legendary biologist Stephen Jay Gould wrote, "This is truly the 'age of bacteria'—as it was in the beginning, is now and ever shall be." Gould wrote this in 1994, before tens of thousands of genomes and environmental metagenomes had been genetically sequenced and analyzed. And there is more to see at the nanoscale: ten times more viruses than there are microbial cells.

From all that we know about Celia Thaxter's life as a gardener, a poet, a painter, and an early conservationist, with all of her insight into living creatures she probably would have been delighted to discover that there is far more depth to nature than she knew. In her book *An Island Garden,* published the same year that she died, Celia noted a white bat that landed on the island one summer. She wrote, "all kinds of strange and remarkable creatures find their way here, and I am surprised at nothing."

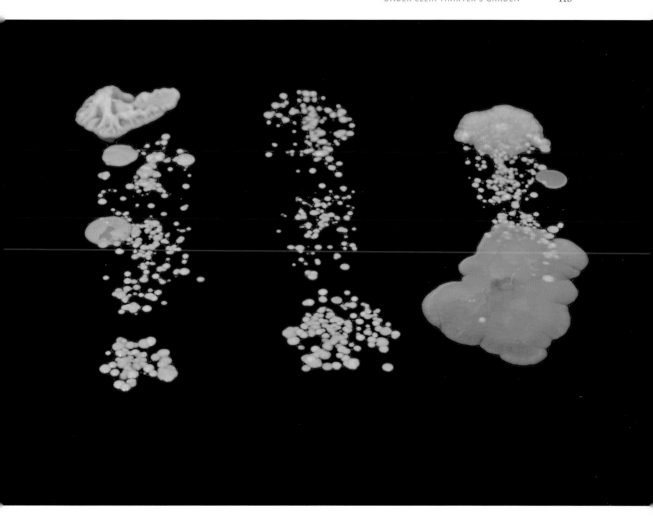

Colonies develop from bacteria left behind by human fingerprints.

4

Intelligent Slime

IT'S NIGHTTIME ON THE TOP FLOOR of a life sciences research center in the city of Boston. There are no voices or footsteps in the hallways, only the machine noise of ventilation fans, shaking cultures, and whistling steam pipes. Hundreds of microbial species grow, cell by cell, inside petri plates stacked in dark incubators. But one microbe in the building is behaving differently. Nearby, a species name is written in fresh ink: *Physarum polycephalum,* the many-headed slime mold.

Inside of the petri plate, a network of bright yellow tubes is moving across the semi-solid agar medium. Each tube pulsates like a vein, expanding to twice its width and contracting to its original size once every minute. The tip of each tube extends into fan-shaped sheets that act like feet. Pulling itself forward, it creeps along centimeter by centimeter. Behind the extensions at the leading edge, tubes connect to form a gooey web.

It appears to be a life-form out of a science-fiction story, an organism teleported from a planet in a distant galaxy. It is, nevertheless, an amoeba: a eukaryotic microbe that evolved right here on Earth over a billion years ago, colonizing land before any plant or animal. However, *Physarum* is not like other microbes. Most of the time, it's not a microbe at all. One day, it might be far too small to see, hidden within leaf litter on the forest floor. The next day, if it encounters enough bacteria or fungi, it becomes a voracious predator, rapidly growing into a macroscopic slime called a plasmodium that can cover an area larger than a square meter. It transitions from invisible flagellated amoeboid cells through the visible feeding and fruit body stages, then back to invisible reproductive spores that can lay dormant and then germinate decades later. Its life cycle is situated directly above and below the limits of the human eye, traversing the micro and macro realms.

As big as it can get, every time we see *Physarum* on a petri plate or on a decaying log, we are looking at a single cell. *Physarum*'s plasmodium is a syncytium: it is made of a continuous cytoplasm. There are no cellular barriers. Instead, *Physarum* plasmodia contain hundreds of thousands of free-flowing nuclei. Parts

Life science research buildings in Boston, Massachusetts.

Petri plates stacked in a micro-biology laboratory.

of the cell might be many centimeters away from each other, but they are con-nected through a single cytoplasm that is held together by a single cytoskeleton. The cytoskeleton works by using the same core cellular components found in human muscle: a structural actin framework and the motor protein myosin. The actin-myosin cytoskeleton is in constant flux. With each rhythmic pulse, it pumps the fluid cytoplasm, distributing nuclei and other cellular contents back and forth through the web of tubes. Along with ostrich eggs, the human egg cell (the ovum), and marine kelp, plasmodial amoebas are some of the largest single cells on the planet.

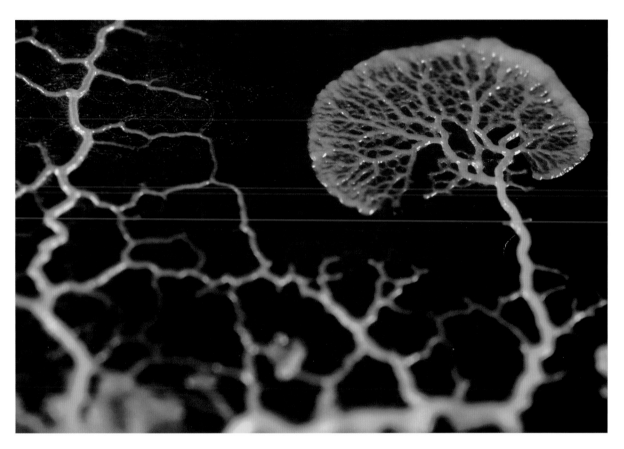

The slime mold network on the petri plate started growing just twenty-four hours earlier when the petri plate was brought into the incubator. At that time, there was nothing much to see on the surface of the growth medium. The micro-biologist who set up the plate took a little piece of a *Physarum* plasmodium from an older plate and smeared it onto a fresh medium with a metal loop. The loop broke the intact plasmodium into thousands of microscopic fragments. With the naked eye, this would have appeared as a faint yellow smudge on the agar sur-face. But if we were to look at the smudge with a microscope, we would see beads of plasmodium beginning to form new cytoplasmic streams.

The slime mold *Physarum polycephalum* forms a gooey web.

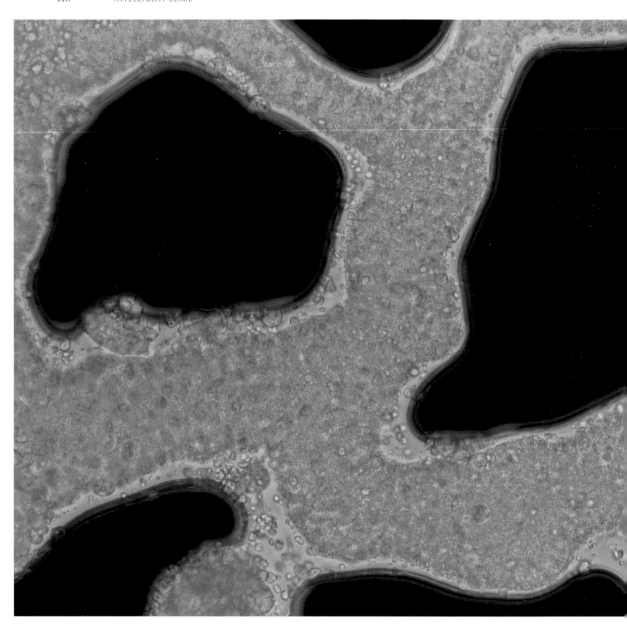

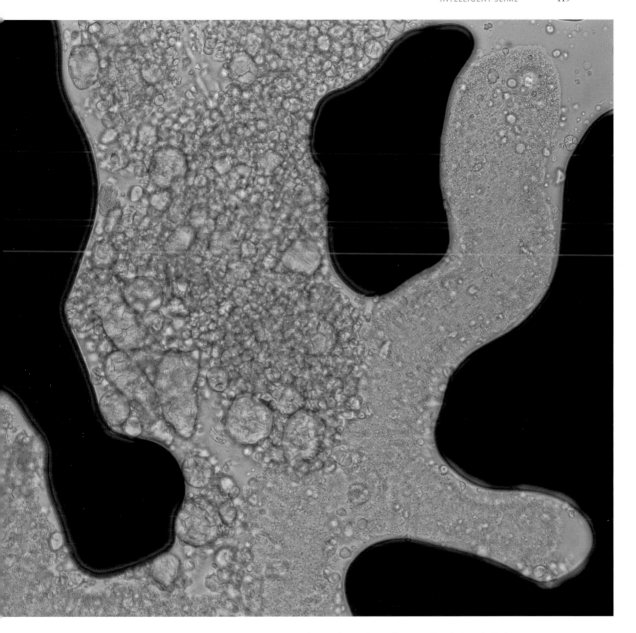

A contiguous cytoplasm with many free nuclei pumps through the web of tubes.

Physarum escapes from its
petri plate.

Over the next few hours, the cell fragments combined with each other, merging and reassembling a syncytial plasmodium. The plasmodium continued to grow, and by the time we first saw it in the incubator, it had consumed almost all of the bacteria added to the medium as a food source. Unlike most microbial species in the laboratory, *Physarum* did not wait around to perish after running out of food. It simply got up and left.

One side of the tube network charged up the wall inside of the petri plate. It used a sensing ability called aerotaxis—movement toward oxygen. This allowed it to find a spot where the seal around the plate was loose and follow a trickle of oxygen. Before long, one extension had reached the top of the plate, pushing against the ceiling of its plastic universe. It squeezed through a crack smaller than the width of a piece of paper. Free, the amoeba rushed down the outside, finding the next petri plate in the stack beneath it and breaking in to explore that environment. Once it made contact with the fresh new medium inside, it began following another chemical trail, this one leading to more food.

Receptors on the surface of the expanding plasmodium recognized chemical attractants released by bacteria on that plate. This caused an immediate response inside of the cell, right under that tiny patch of the cell's surface. The actin-myosin cytoskeleton pumped faster, resulting in increased cytoplasmic streaming toward that local area. Its actions excited the area of the cell around it, and that patch the next, sending a wave of information from the tip of the vein up through the cell and into the network on the plate upstairs. So long as the attractant is still there, the process continues through a positive feedback loop, causing the entire network to abandon the original petri plate. Inside the new plate, the plasmodium expands in all directions. In every location that a bacterial colony is found, local patches of the plasmodium surface and nearby area of the cytoplasm function as a cohesive acellular entity known as a biochemical oscillator. Oscillators send out attractive signals, causing that part of the network to expand over time. In contrast, areas of the tube network where no food or favorable conditions are found contract. The ultimate result is a network that connects all available food sources through the shortest possible routes, plus one backup connection in case the first one is severed.

Physarum mapped its environment. Wherever it had not found food, it retreated, leaving behind a translucent film. Like footprints on a trail, this extracellular slime let *Physarum* know where it had already searched, ensuring that it would not waste time and energy doubling back. It had some sense of the dimensions of the plate; it knew where all of the food sources were, and it was able to efficiently distribute nutrients throughout the giant plasmodial cell. It had developed a kind of navigational memory. The memories were not stored internally in the way a human brain stores memories or a computer stores information on hard drives. *Physarum* stored and accessed memories within the environment itself.

So when we look at a *Physarum* network on a petri plate from above, we know quite a bit about where it is headed and where it has been. See the wider webs at the expanding front and the thickest veins leading to those fronts? That's where the plasmodial cell is actively exploring. It is where *Physarum* is moving and sensing. If one of these webs looks particularly thick, there is probably a food source in that direction. The cytoplasm and cell surface in that local area are lit up like a biochemical beacon, communicating the location of the possible food source across the entire network. If we look back in a few minutes, that yellow blob will probably be the blob that has expanded the fastest. And the translucent sections of the network? That's where *Physarum* has been. We know that as long as there is more open space to explore, it will not return to that part of the plate.

When the microbiologist returned to the incubator to retrieve her *Physarum* culture, the plate was empty. We can imagine the look of fear on her face as she thought, "Where is the slime?" She checked the other plates in the stack. By then, her *Physarum* culture wasn't on the next plate down. All of the bacteria on that plate were gone, leaving only a telltale trail of slime. On the next plate, nothing. Again, the bacteria had been eaten. Finally, she found the active slime, caught red-handed four plates down and about to conquer the next one. "What kind of monster microbe is this?" she thought.

The biochemical oscillators within plasmodial amoeboid cells are hyper-interactive biological units. They are the building blocks of a biological system with many of the same properties of a multicellular animal. This unique system evolved because the survival and reproduction of *Physarum* depend upon how well individual amoebas work together. When fragmented, plasmodial cells attract each other like living magnets. They rapidly assemble from microscopic parts into macroscopic bodies and forage through much greater areas than any single microscopic amoeba could. Even without a nervous system, they make decisions and solve problems. They find the shortest path through a maze. Researchers have even used *Physarum* to simulate human networks such as subway systems. They have placed food sources at many discrete locations and observed how *Physarum* creates routes between them. In every challenge, the plasmodial cell remaps the routes that human engineers have designed, only a bit more efficiently. And they do this through a process of self-organization. There is no single part within a plasmodium that organizes and directs the other parts of the cell. All of the biochemical oscillator units are identical and interchangeable. Any tiny area of the slime network might sense the next food source and send the signals that direct future movements. Plasmodial amoebas are the masters of these coordinated or collective behaviors: highly synchronized, highly redundant, and highly interactive.

At the same time, when we take almost any microbe from the natural environment and grow it in petri plates, cellular or acellular, we see similar propensities for interaction. And from these interactions, we see emergent properties: multicellular structures and collective behaviors with synergistic qualities greater than the sum of the parts.

It's tempting to think of a bacterial cell, for example, as a single-celled organism that swims around in its environment. Let's take the example of one bacterium, *Bacillus subtilis*. In nature, *Bacillus subtilis* is ubiquitous. We can easily find it in damp forests and arid wastelands. In studying *Bacillus subtilis* we move away from the strange biology of biochemical oscillators of slime molds and find the much more common organizational unit of most bacterial and archaeal species: a small cell. Bacteria from the genus *Bacillus* were first studied in the late nineteenth century by the German scientists Ferdinand Cohn and Robert Koch. Their detailed description provided glimpses of the complex lifestyles of these bacteria. *Bacillus subtilis* cells are small; hundreds of thousands of them would fit in the eye of a sewing needle. But the environments that *Bacillus subtilis* lives in are structured. If ever there is a single *Bacillus* cell on the side of a soil particle and if conditions are favorable for growth, the cells will divide, and there will soon be hundreds of *Bacillus* cells in that microscopic location. These larger groups of cells are called microcolonies.

If we take *Bacillus subtilis* isolated from the wild and grow it in the laboratory on an agar surface or within a liquid growth medium in a beaker, microcolonies grow to become larger, more complex structures called biofilms. A biofilm is more than a large pile of millions of cells. As cells divide within a microcolony, they release a variety of molecules that form a gelatinous extracellular matrix. The extracellular substances are functionally similar to the connective tissue that holds animal cells together in tissues, except that instead of the molecule collagen, bacterial extracellular matrices are made from sugary exopolysaccharides, sticky amyloid proteins, and extracellular DNA. Just one cell can grow to form a variety of macroscopic, three-dimensional structures made of many millions of cells. This way of living allows microbes to stay put and form a community in an environment where there may be a stable food source, such as a nutritious medium in a petri plate.

Without knowing the exact dynamics or extracellular matrix molecules of biofilms, the basic concept of biofilms has been known since the 1930s and 1940s through the work of Arthur Henrici and Claude Zobell. Henrici discovered biofilms when he left clean glass slides in freshwater lakes. When he removed the slides, he noticed that cells within the water formed a "sheath of gum which also serves to fasten the colony to glass." In 1933, Henrici wrote that "for the most part the water bacteria are not free-floating organisms, but grow on submerged surfaces; they are of the benthos rather than the plankton."

124

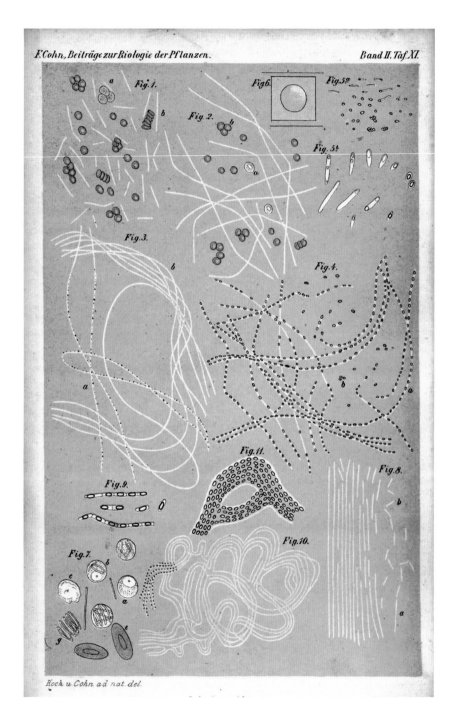

Band II. Taf. XI.

Koch u. Cohn ad nat. del.

A first look at the biology of *Bacillus* species, drawn by Ferdinand Cohn and Robert Koch in 1877. Cohn was working with the harmless species *Bacillus subtilis* and Koch with the deadly species *Bacillus anthracis*.

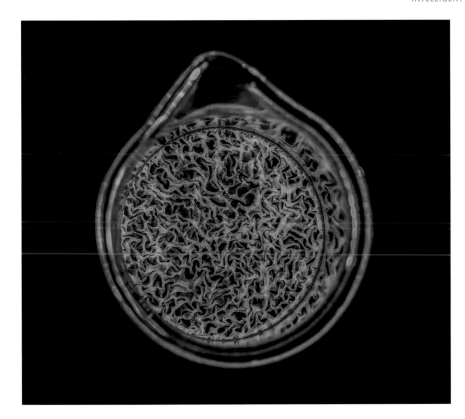

A floating pellicle biofilm of *Bacillus subtilis* forms within a beaker.

A decade later in 1944, Zobell published diagrams of bacterial cells surrounded by what he called "absorbed nutrients" that are instantly recognized as what we call biofilms today.

Most bacterial and archaeal cells in nature grow in structured communities on surfaces. Right now, there are biofilms living on almost every surface on Earth. They occur between solid materials and air, such as the biological crusts that form on desert sandstone, on our own teeth, on stone monuments, and on salami rinds. Biofilms grow submerged in aquatic habitats like rivers, lakes, and ponds, in pipes within our buildings, and on the hulls of seafaring ships. Within submerged biofilms, microscopic channels surround clusters of cells, acting like a primitive circulatory system where nutrients and waste products flow through the biofilm.

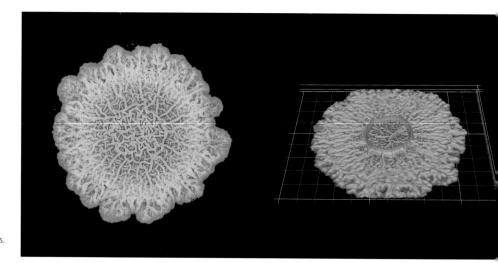

The three-dimensional structure of *Bacillus subtilis* colony biofilms.

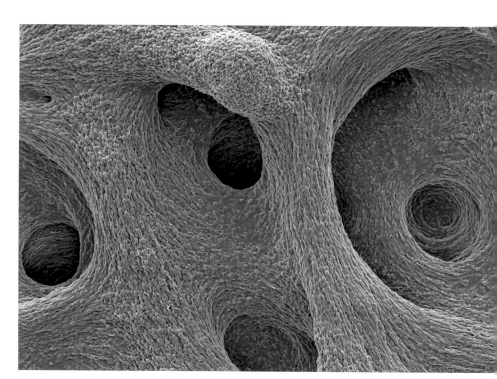

Cells within *Bacillus* biofilms linked by a common extracellular matrix.

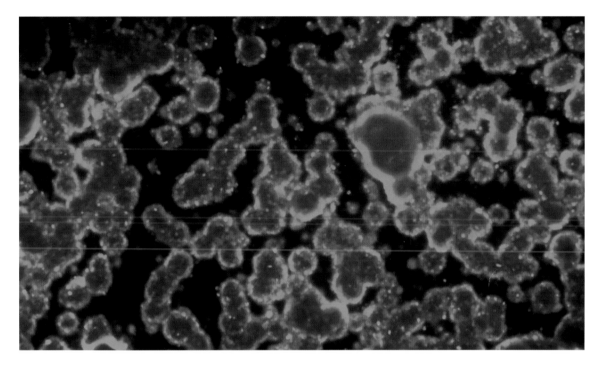

Water channels allow flow throughout a submerged biofilm. This biofilm was formed by the halophilic archaeon *Haloferax volcanii.*

Biofilms are not the only lifestyle that bacteria use to adapt to their environment. Sometimes, the opposite life cycle promotes survival: sometimes cells need to migrate. *Bacillus subtilis* can swarm over surfaces in large multicellular groups. To do this, groups of cells secrete a soapy molecule called surfactin that reduces the surface tension of water. Surfactin moves out from a swarm of bacteria like a chemical moat, altering the environment and enabling the rapid migration of specialized hyper-swimmer cells. Different species produce a range of swarming behaviors and patterns on two-dimensional surfaces. *Bacillus subtilis* swarms split into branching dendrites.

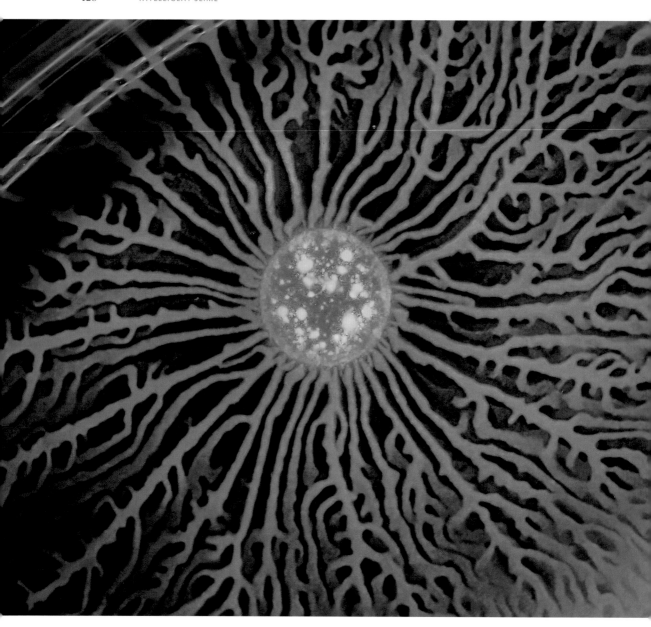

Bacillus subtilis swarms as one, forming a dendritic pattern as it expands.

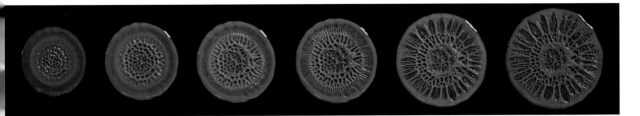

And even though an infinitesimally small fraction of the microbes on Earth cause disease in humans, those species that are harmful to us also build biofilms and swarm together. *Pseudomonas aeruginosa* cells develop into ornate wrinkled biofilms in the laboratory. These wrinkles increase surface area and therefore increase access to oxygen within the thick layer of cells within the biofilm. It is also an opportunistic pathogen. This means that when the conditions are right, *Pseudomonas aeruginosa* can shift from a harmless bacterium found in many environments to a pathogen that causes infections within burn wounds in humans. *Proteus mirabilis* is another bacterium that is an opportunistic pathogen; it engages in one of the most unique and well-studied swarming behaviors, producing a characteristic terraced pattern as it moves across surfaces in the laboratory. There are thousands of specialized swarmer cells at the leading edge of every ring within *Proteus mirabilis* swarms, each at least twenty times longer than nonswarmer cells and each with at least twenty times more flagella. These swarmer cells gather and move forward rapidly as one multicellular raft.

Much like the enhanced foraging and decision-making ability of the plasmodium stage in *Physarum,* biofilms and migrating groups have emergent properties unavailable to individual cells. And for biofilms and swarms, they amount to a Pandora's box of new functional abilities in bacteria and archaea. Molecules diffuse slowly into the extracellular matrix and are sometimes directly inactivated by matrix components, protecting cells within a biofilm from toxic chemicals. In the context of infection, this emergent property of biofilms becomes quite important, because the toxic chemicals in question might be antibiotics administered by a physician. The protective effect of the matrix may render the bacteria in the biofilm resistant to the antibiotics. The matrix keeps cells hydrated and prevents them from drying up if conditions take a turn for the worse. The matrix can also absorb nutrients from the environment, and it contains enzymes that

The bacterium *Pseudomonas aeruginosa* develops as a colony biofilm on solid growth medium over several days.

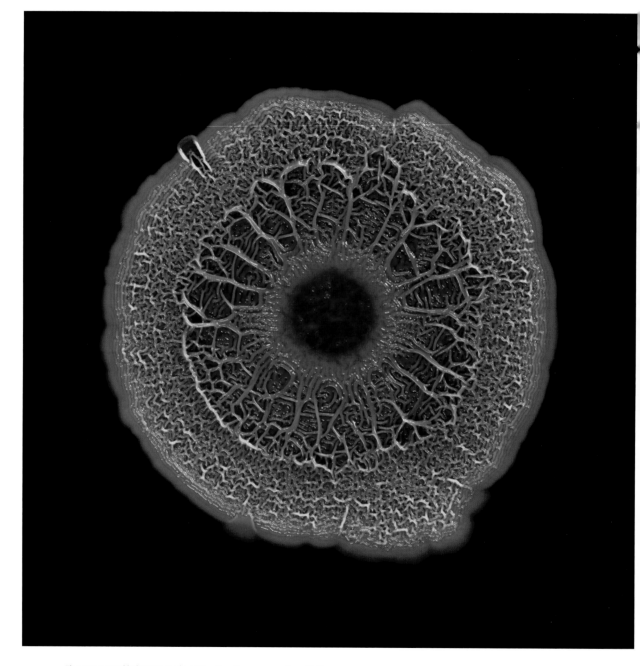

The ornate, wrinkled structure of a mature *Pseudomonas aeruginosa* biofilm.

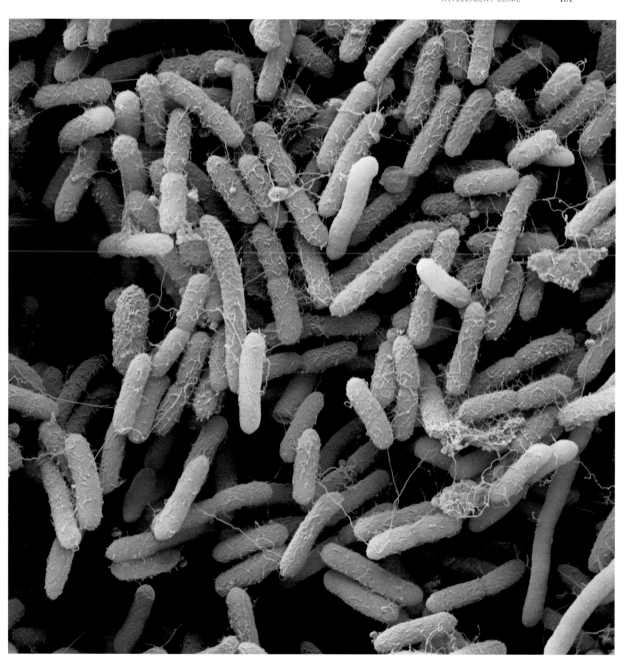

Extracellular matrix and pili connect cells in a *Pseudomonas aeruginosa* biofilm.

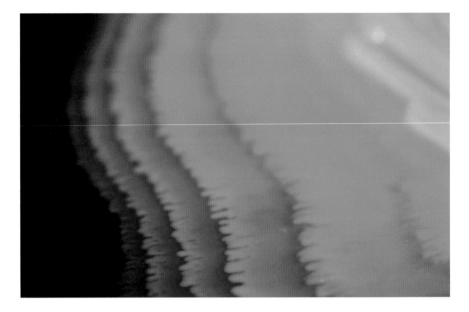

Proteus mirabilis forms terraced patterns as it swarms across a surface. This strain of *P. mirabilis* expresses green fluorescent protein.

break down complex materials outside of the cell, functioning as a shared extracellular digestive system.

The extracellular matrix itself also becomes a habitat for other microbial species. The physical framework of the biofilm creates a mixture of conditions at the microscopic scale. There are patches with more or less oxygen and with more or fewer nutrients and waste products. These different areas form different niches for microbes to exploit. Different microbes adapt to live at different places within a biofilm much as animals and plants adapt to specific zones in a rainforest, from the innermost regions to the towering canopies. On an even smaller level, different species within biofilms form micro-clusters where they exchange metabolic products. Synergistic consortia arise in pockets of the biofilm as species naturally aggregate based on their mutual metabolic needs. One species uses the metabolic by-products of another species and vice versa: a process called syntrophy.

Proximity between cells in a biofilm also enables cells of the same species to communicate with each other and develop more complex structures and behaviors. For *Bacillus subtilis* and other bacteria, communication occurs through a chemical language. This language's lexicon is composed of small molecules and peptides produced by cells. The chemical words are released into the extracel-

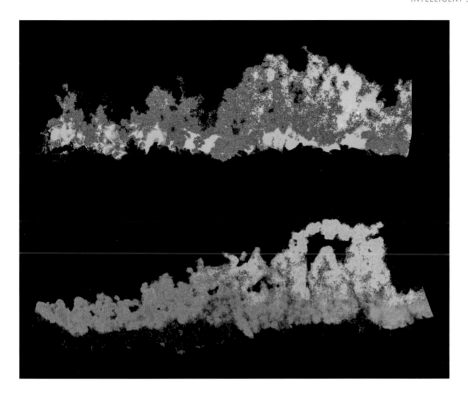

Different functions occur at different places inside a *Bacillus subtilis* biofilm. Red cells produce the extracellular matrix; blue cells are motile cells, and yellow cells are spore-formers.

lular environment, where they are recognized by other cells. Chemical communication can be used to coordinate the expression of particular genes in many cells simultaneously, a phenomenon known as quorum sensing. Communication can also be physical, mediated by a variety of mechanisms that make contact from one cell to another and permit the flow of genetic and chemical information. Communication between microbes can even be electrical, involving the transfer of electrons between cells.

In *Bacillus subtilis,* patterns of chemical communication lead to the development of many different functional cell types. Some cells have flagella rendering them motile, free to move around. Some cells produce the extracellular matrix; some form spores, and some release the slippery surfactin molecules. New functions emerge from interactions between these cell types. If we cut a *Bacillus subtilis* colony biofilm in half to see a cross section and label the cells in the biofilm with fluorescent proteins of different colors based on their varying functions, we can see that different activities are carried out in different parts of the biofilm. Yellow cells form spores at the top where they are most effectively released into the environment; blue motile cells migrate to the bottom, and red cells that produce extracellular matrix are found through the middle section of the biofilm.

If we look at a *Bacillus subtilis* colony from above and examine the edge, a similar kind of division of labor occurs between cells that produce extracellular matrix and cells that produce surfactin. This example is not based on the distribution of spores, but instead drives a form of collective surface migration called sliding. Here the matrix-producing cells are visualized with green fluorescent protein; they form long chains that migrate alongside red surfactin-producing cells, forming multicellular loops. The undulating multicolor patterns that emerge are reminiscent of van Gogh's paintings and are thus called van Gogh bundles. These structures increase the fitness of the species, allowing cells to reach resources well beyond the edge of the colony.

Local interactions between individual units in large populations are the driving force of all collective behaviors. If it is a honey bee colony, collective decisions emerge from acoustic messages transmitted between honey bees. If it is a *Physarum* plasmodium, foraging behaviors emerge from biochemical signals that pulse across the cell. And if it is the protective qualities of a bacterial biofilm, cells interact within the extracellular matrix.

Myxobacteria are one of the most interactive and cooperative species of bacteria. They are so cooperative that they form structures that look like fungi rather than bacteria. Roland Thaxter first discovered myxobacteria in the late 1800s during his walks in the Maine woods. He eventually determined that they were bacteria and not fungi. He described several species of myxobacteria as colonial bacteria and drew them as they appeared to swarm together, "possessing a power of slow locomotion and secreting as they multiply a firm gelatinous base which connects the colony as a whole." Today we can observe the development of this locomotion as a thin leading edge of a spreading colony using time-lapse photography, which enables us to study it in great detail. Thaxter, with fewer resources, had no concept of the fine-scale local interactions that led to the social behaviors he saw. These molecular interactions are the signals and cell surface proteins that lie in the seemingly empty spaces between swarming cells in his

(Opposite) Sliding motility: another example of division of labor in *Bacillus* communities. Red cells are producing slippery surfactant molecules and green matrix-producing cells form structures called van Gogh bundles.

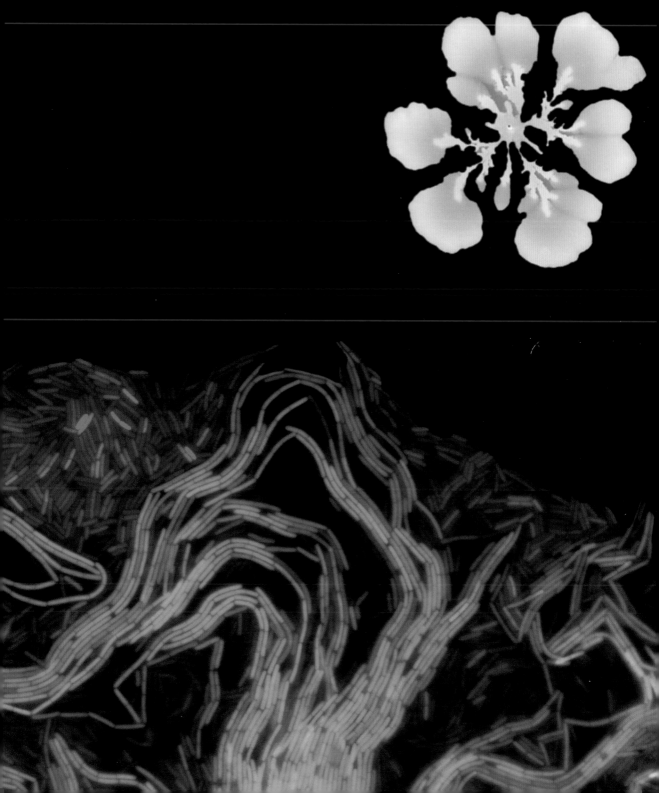

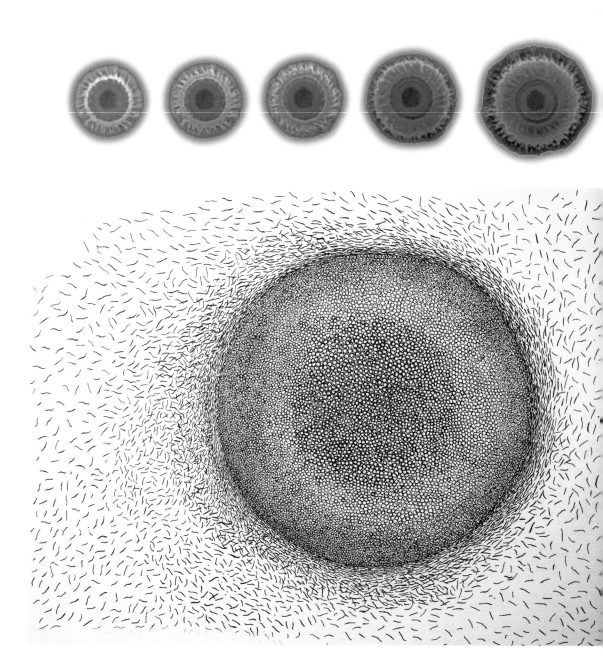

drawings. Because myxobacteria live in highly variable soil habitats and grow slowly, cells in natural populations exist at various ages and physical conditions. But like plasmodial amoebas, myxobacteria must come together to form multicellular fruiting body structures in order to complete their life cycle. This has driven the evolution of a mechanism of social interaction in which cells temporarily merge with each other during swarming and exchange sections of the cell membrane. As they do so, damaged cells become rejuvenated by healthier cells. It is a way to bring cells that otherwise could not help in the formation of a multicellular structure back into working order for the benefit of the population as a whole.

Most of the knowledge we have gained regarding microbial behaviors has come from studies of individual species, such as *Bacillus subtilis,* grown as a pure culture in the laboratory. While we must study species on their own to understand them at the molecular level, we have the ultimate goal of reconciling this knowledge in the context of the microbes' natural ecology. And in nature, there is always a thin line between social cooperation and competition.

In microbes, the more social a behavior, such as the rejuvenation of myxobacteria cells, the more important it is that the cells engaging in that activity are close genetic relatives. A myxobacterial cell cannot trade parts of its membrane with just any other cell. After all, this would lead to no advantage for that donor cell: no hope that this would help the population to develop into a multicellular fruiting body. All molecules, whether traded between cell membranes or secreted into the environment, cost energy to produce; they are a kind of currency or public good of the cell. It's the same reason that we keep most of our paycheck for ourselves and our families. Most people can afford some charity, but if we give away too much money, we will reach a point where we can no longer survive. This means that in order to be cooperative, species must also compete with cells from other species and even with cells from other strains within the same species.

Mechanisms of competition have evolved in the microbial world in parallel to the most advanced cooperative behaviors. These are strategies for keeping competing species out of the immediate area and ensuring that public goods benefit the species that produces them. *Bacillus subtilis* can discriminate between its close relatives and distant ones; that is, it distinguishes kin from nonrelatives. Instead of recognizing kin by physical appearance as people do, *Bacillus* uses a molecular mechanism to ensure cooperation with close genetic relatives.

The myxobacterium *Myxococcus xanthus* spreads across a surface over several days.

Multicellular movements of a species from the genus *Myxococcus* as drawn by Roland Thaxter in 1892.

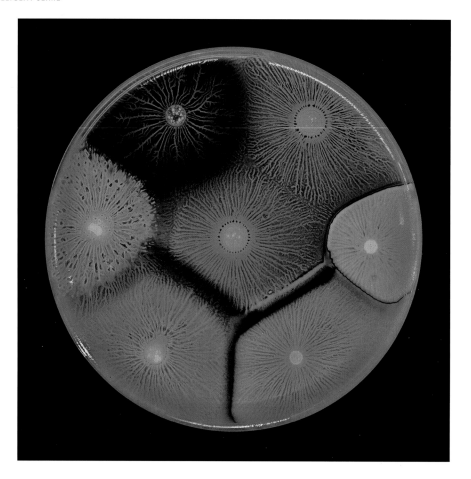

Swarms of different *Bacillus subtilis* strains either merge or form boundaries.

If two populations of cells of the same strain swarm into each other, they merge. But if two populations from different strains swarm into each other, they form a boundary. The boundary is a battlefront, where cells from each group have lost their lives in an intense chemical war zone. Other bacterial species, among them *Proteus mirabilis,* form a molecular spear that injects nearby cells with a poison. If that nearby cell is closely related, it is safe: it has the gene encoding the antidote. If the nearby cell is not a close relative, the toxin will break the cell membrane, causing the cell to burst apart. Here again, when migrating cells from different lineages meet, they create a clear zone between them.

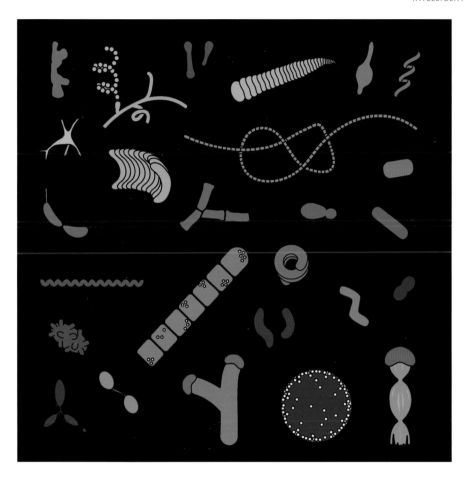

This social and multicellular biology seen at play within and between microbial groups indicates that the concept of unicellular and multicellular organisms as two schemes of biological organization is overly simplified. In some cases cells must be physically linked together to complete an organism's life cycle, a state called obligate multicellularity. This kind of life cycle has evolved many times in independent branches of the tree of life. But then there are all of these other optional forms of multicellularity in microbes that we've seen, such as in biofilms and microbial mats. We call these facultative; they are multicellular forms that are not required for reproduction but are relevant to the ecology and

The multitude of microbial cell morphologies.

evolution of those species. Variations in cellular morphology alone have led to a bizarre array of multicellular forms in bacteria and archaea, from straight filaments to branched filaments, bulbs, stalks, spindles, and spirals. Some of these are easier to see under the microscope—biofilms, filaments, and branched cellular arrangements—but there are many interactions that are more difficult to spot.

Consider the case of *Prochlorococcus*: a picocyanobacterium that forms tiny spherical cells. It is also one of the most abundant species. There are an estimated 10^{27} *Prochlorococcus* cells on Earth—more cells than there are stars in the known universe. Despite tremendous variation in water temperature, sunlight intensity, and water chemistry, this bacterial species is able to proliferate in the top layers of the ocean over the entire planet. It does not live within biofilms. It seems to defy any form of a community lifestyle.

Yet the ability of *Prochlorococcus* to spread across the planet is another emergent property of the collective population. *Prochlorococcus* has a small genome containing only around 2,000 genes. Only about half of those genes are shared between all cells in the global *Prochlorococcus* population. This is the core genome: it encodes the cell's most essential housekeeping genes for basic cell functions. The other 1,000 genes in the genome can vary among cells and are referred to as accessory genes. The conglomerate of the core genome plus all the accessory genes is known as the pangenome. The pangenome is an assortment of genes that makes up a pool of an estimated 57,000 genes in total, distributed across the global *Prochlorococcus* population. The accessory genes allow different subpopulations of cells to adapt to different niches in the ocean. The process that generated this diversity of accessory genes, the same process that maintains the diversity today, is an active exchange of genetic information between individual cells. Every time a bacteriophage virus infects a *Prochlorococcus* cell, there is good chance its progeny will carry a section of that cell's genome into the next cell that is infected. These nanoscopic bacteriophages are less than one-tenth the size of *Prochlorococcus* cells. They have sharply shaped heads that contain their genomes, long cylindrical sheaths, and feetlike tail fibers used to attach to their bacterial hosts. At the rate of millions of infections per minute in every swath of ocean water, bacteriophages shuttle genetic information through the intercellular space. The bacteriophage genomes are selfish genetic elements, out to replicate themselves, but nevertheless they have shaped the *Prochlorococcus*

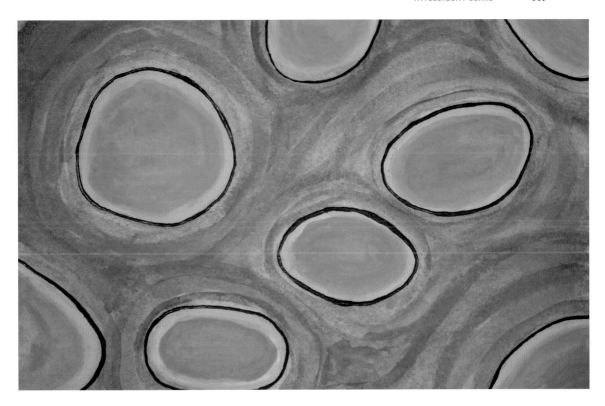

An artistic concept of individual *Prochlorococcus* cells that live in the ocean.

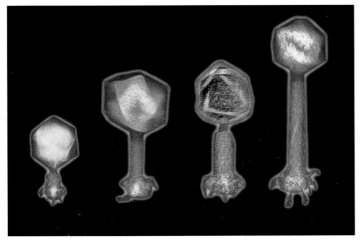

Cyanophage viruses that infect *Prochlorococcus* cells. The first phage displays a shorter, contracted tail; the second, of the same type, an uncontracted tail. The third and fourth phage are of a different type, also showing a contracted tail and an uncontracted tail.

genome. In addition to gene exchange through viruses, *Prochlorococcus* cells release small structures called extracellular vesicles that also carry DNA and other biomolecules between cells.

Prochlorococcus cells are minimal cellular units with a streamlined genome and an efficient surface-to-volume ratio. But the long-term evolutionary success of the species as a whole relies on the sum content of genes held in collective *Prochlorococcus* populations. There are even important genes outside of the pangenome, stored off-site in the bacteriophage population. *Prochlorococcus* populations are genetically divisible into dynamic ecological subspecies called federations or meta-populations, depending on which metaphor we prefer. Without the pangenome of those federations and of wider, global groups of federations as a repository of genetic information, without the unseen interactions between bacteriophages, extracellular vesicles, and cells, it could not be the most abundant photosynthetic species in the ocean.

Prochlorococcus and many other microbial species teach us that multicellular biological phenomena don't occur only between cells that are physically linked together. They also occur between cells that are separated in time and in space. If we label all the *Prochlorococcus* cells in seawater and look at them under the microscope, we see what appears to be single cells divided in space. However, any cell that exists as a true solitary biological entity will be the most ineffective organism imaginable; it cannot persist to recopy its genetic material in the next generation. Even the most basic reproduction through binary fission links cells within a population. Every "free-living" *Prochlorococcus* cell was once physically connected to another cell; it emerged from the biochemistry of that cell, and that cell came from the cell before it. At any given moment, the trillions of living *Prochlorococcus* cells are united by an unbroken chain of cellular divisions through time, even if every cell is released into the ocean as soon as it divides, never to re-encounter its sibling cell again.

Prochlorococcus has also evolved to interact with and trade molecules with other free-living species. Most cyanobacteria have a gene encoding a catalase-peroxidase enzyme that is released outside the cell. Without this enzyme, cyanobacteria cannot survive when exposed to hydrogen peroxide, a deadly chemical in the environment. *Prochlorococcus* lives in the same environments with hydrogen peroxide, but the gene for catalase is missing from its genome. So how does it survive?

The missing catalase gene is explained by an evolutionary theory called the Black Queen Hypothesis. Rather than producing its own catalase, *Prochlorococcus* benefits from copies of this enzyme produced by other species living nearby in the ocean. In this case, a bacterium called *Candidatus Pelagibacter*, which lives side by side with *Prochlorococcus*, produces the catalase. The hypothesis is based on the principle that energy is required to manufacture enzymes inside of the cell. So, because there is always enough catalase around, there is no fitness benefit for *Prochlorococcus* to keep its own catalase gene. In this case, the catalase is like the queen of spades in a game of Hearts: it's the card to be avoided because it carries the most points at the end of the game. And *Candidatus Pelagibacter* is currently holding the card. There are other examples of such interdependent co-evolution between cells of different species within microbial communities across the planet. Even *Candidatus Pelagibacter*, though it has the queen of spades in the context of the catalase gene, could be the beneficiary of the same Black Queen interaction with respect to other genes that it has lost over time. There's an unseen network of metabolic interactions between cells in nature, with no respect to the species names we assign or our concept of what a single organism is.

The lesson from these many microbial systems is clear: there is no negative space in between cells in the microbial world. Rather, cells are interactive biological units by definition. Whatever the ecological setting may be, microbial species exist within a sea of interactions. At minimum, even if they are not physically linked together in a biofilm or microbial mat, they are surrounded by viruses, extracellular DNA, extracellular vesicles, and small molecules that transfer and turn over genetic information, nutrients, and communication signals among the community.

Emergent properties of the smallest forms of life ripple up through larger size scales. *Prochlorococcus* is the smallest photosynthetic organism found so far on Earth. *Candidatus Pelagibacter* is the smallest free-living heterotrophic species in the ocean. But *Prochlorococcus*, as it relies each moment on enzymes produced by *Candidatus Pelagibacter*, impacts the global carbon cycle. If all photosynthetic picophytoplankton suddenly died, carbon dioxide would rapidly accumulate as a greenhouse gas in the atmosphere, with major effects on the global climate. The even smaller *Candidatus Pelagibacter* cells, each half a micrometer wide, and the nanometer-scale bacteriophages that infect *Prochlorococcus* also

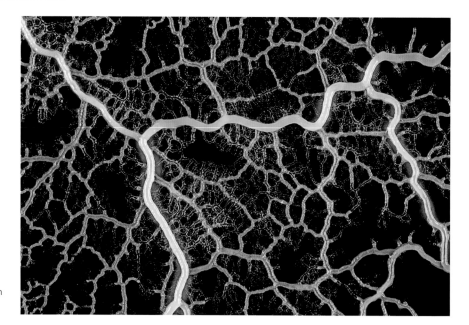

A *Physarum* network as seen from above.

The networked roads and lights of the city of Boston as seen from a satellite.

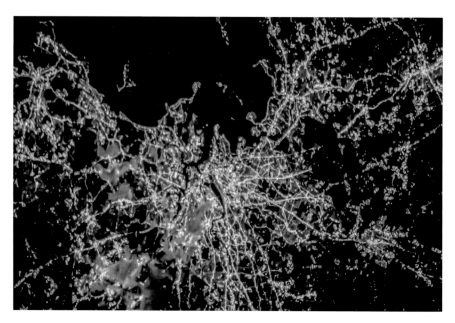

influence the climate of a planet 13,000 kilometers in diameter. And yet *emergence,* the self-organized process through which greater effects are produced by many small entities, is something that we do not fully understand. A natural phenomenon, it should be just another area of inquiry, solvable through hypothesis and testing. Like dark matter and dark energy in physics, we are just beginning to make accurate predictions. It is a problem for the integrative science of the future.

The microbiologist studying the semi-intelligent slime mold in Boston carries her own microbiome of bacteria, archaea, fungi, and viruses, interacting with the cells of her immune system and keeping her healthy every day. Her consciousness is an emergent property of the network of neurons in her brain. She lives in a society. She is part of the most interactive, most intelligent species on Earth. The building where she works is a product of human consciousness. A satellite image of the city of Boston at night shows a network built by human collective behavior that began in the colonial period of American history, much like the fundamental principles that explain the movements of *Physarum* as it searches for food.

Ernst Haeckel's illustration of twenty different macroscopic slime molds.

Emergent structures and behaviors of microbes challenge us to untangle local interactions between cells, between proteins in those cells, and between individual amino acids within those proteins. It's razor-sharp reductionism. We need the kind of experts that the biochemist Erwin Chargaff called the "molecular podiatrists" of the "fifteenth foot of the centipede." Simultaneously, we must integrate this knowledge with patterns and processes that unfold before our eyes: with the slime molds and other natural phenomena that the nineteenth-century polymath Ernst Haeckel illustrated masterfully in *Art Forms in Nature.* We need both views, of the microscopic and the macroscopic, of the interactors and the collective, to understand the microbial world and its impacts within the rest of the biosphere.

5

Tales of Symbiosis

IT WAS A TIME OF TRIBAL WARFARE in the Amazon rainforest of northeastern Peru. The year was 1929, and two villages of indigenous peoples were fighting over land. Meanwhile, two colonies of leaf-cutter ants from the tribe Attini were in battle on the rainforest floor, also over territory. On the ant battlefield, a soldier's leg was severed. Unfazed, she grabbed her opponent by the neck, and with one sharp squeeze, removed the head. Three centimeters away, her comrade-in-arms was pinned. She climbed on the back of the enemy and decapitated her too, another strike against the enemy. But her efforts were futile. Her tribe was losing.

Just as they were completely encircled by a new wave of enemy soldiers, a giant hand came down from the heavens. Soldiers of the opposing ant tribe began disappearing, one after the other, some with limbs of their opponents still hanging from their jaws. They were saved.

Far above the ants, a medicine woman was rushing to collect them. She plucked as many of the large soldiers as she could, oblivious to the insect war. Holding a jarful of snapping, wriggling ants, she ran back to her village.

When she arrived, the men had returned from their own brutal battle, suffering from gruesome cuts by bush knives and machetes. The survivors lined up before the medicine woman. As she began her treatment, a scientist from the Chicago Field Museum named Llewelyn Williams, passing through the area on a botanical expedition, watched the scene unfold. The medicine woman took an ant from the jar and carefully directed the biting mandibles to clench and pull the man's laceration together. Williams recorded the next step of the surgery in his field journal: the medicine woman twisted off the body of the ant and "the lifeless head remains with its death grip on the skin until the wound is healed." Sometimes, he noted, warriors have "half a dozen of these ants' heads holding a large wound closed."

The wounds healed remarkably well with ant heads as natural stitches. How did they work so well, despite the fact that they were far from sterile? The secret lies back in the wilderness where the ants were collected by the medicine woman, deep within the territory of the leaf-cutter colony.

A soldier leaf-cutter ant defends smaller worker ants.

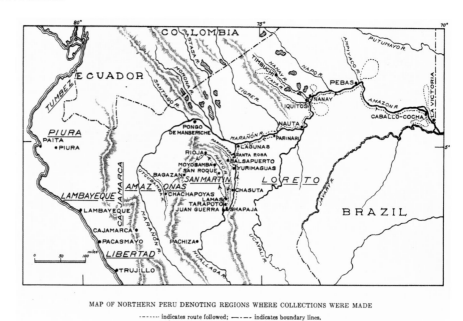

MAP OF NORTHERN PERU DENOTING REGIONS WHERE COLLECTIONS WERE MADE
------- indicates route followed; —-—- indicates boundary lines.

Map of Llewelyn Williams's Botanical Field Expedition in 1929. It was during this trip that he witnessed the use of ants as stitches by indigenous peoples.

Leaf-cutters were first classified in 1758 by Carl Linnaeus, the Swedish botanist and zoologist responsible for the modern system of classifying species. Different species of these ants inhabit various terrains of the Americas, from Texas to Argentina. Their ambitious leaf-cutting operations have intrigued explorers for centuries. At any given time, thousands of specialized female worker ants forage for leaves from the lowest bushes to the highest treetops, with mandibles specifically adapted to cut plant matter into pieces. The ants hold the leaf fragments above their heads like tiny parasols, earning them the common name *parasol ants*. If we were to look down at the area around a colony, we would see an industrial-scale chain of assembly lines that all lead into the underground nest. Up to 20 percent of all leaves in the tropical rainforests are harvested by leaf-cutter ants. But as impressive as the ants' above-ground operations are, the most fascinating aspects of leaf-cutter ant biology occur underground, after the leaves are brought into the nest.

One of the first people to explore the underground nests of leaf-cutter ant colonies was the German geologist and naturalist Thomas Belt. Thomas was leading the operation of a mine in Central America in the late 1800s. There he carefully observed and recorded the features of the natural settings that surrounded

(Left) A worker leaf-cutter ant, *Atta colombica*, uses specialized mandibles to cut sections of leaf.

(Below) Workers forage from the forest floor all way up to the rainforest canopy.

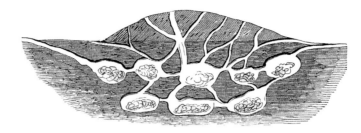

All scavenged leaves are brought into an underground network of chambers in the ants' nest, as shown in this illustration from *The Naturalist in Nicaragua*.

him. From ants to jaguars to ancient stone sculptures, from mental processes in wasps to the origin of cyclones, he captured it all in one story: *The Naturalist in Nicaragua*. Thomas was first acquainted with the leaf-cutters when his garden was eaten by them. They ate most of his banana, orange, and mango trees, and when he followed the trail of leaves, he found a small leaf-cutter ant colony about one hundred yards away, at the edge of the forest. He tried plugging up the entrance, but the ants quickly built another. He dug down to destroy the colony but realized that it was so large his efforts would be pointless.

Later on, a large colony was unintentionally excavated during their mining operations such that the ants' "underground workings were exposed to observation." Thomas sketched the nests, which were round chambers, each about the size of a human head, all joined by a network of tunnels. These chambers were filled with a loose, spongy material. He collected some of the material and examined it closely, finding that it was a mixture of small bits of leaves that had withered to a brown color, "connected together by a minute white fungus." He found the fungus in every chamber that he opened and noticed that the ants went to all ends to preserve it when their nests were disturbed, even carrying it off in chunks to establish another colony.

Thomas had figured out one of the most fundamental aspects of leaf-cutter ant biology. Many naturalists had studied them; they were curiosities for most who traveled to the tropics. These ants were the enemies of many human gardeners, but no one knew exactly what the ants were using the leaves for. Some observers thought the leaves were used directly as food and others thought they were used as structural supports or roofing within the nest. Through careful observation, Thomas determined that "the real use they make of them is as a manure . . . they are, in reality, mushroom growers and eaters."

Thomas was absolutely correct: leaf-cutter ants are the quintessential farmers of the animal kingdom. They made the switch from a hunter-gatherer lifestyle to an industrial-scale agricultural system capable of sustaining millions of individuals. Modern genetic analyses and fossilized ants in amber indicate this shift to farming occurred sometime at the end of the Cretaceous period, 65 million years ago. There were no humans. Our living ancestors at the time were mammals that ran along tree branches and around the ant colonies on the forest floor, looking something like modern squirrels and shrews.

Once the workers bring the leaf pieces into the nest, the fungus farming begins in earnest. Smaller gardener ants within the colony chew up all plant material and carefully arrange it within chambers in order to cultivate a particular type of fungus called *Leucocoprinus*. This fungus happens to be in the same family as some of our favorite edible mushrooms. The fungus has enzymes that break down the leaves, allowing the ants to indirectly access the nutrients stored in complex molecules within the plant tissue. The gardens act as an external digestive system for the colony.

Leaf-cutter ants and the fungi in their gardens lived together for so long that they became dependent on each other for survival: a symbiotic relationship known as mutual dependence. Some *Leucocoprinus* species are not found in nature outside of ant colonies, and leaf-cutter ants cannot live without fungal gardens. Garden fungus is the only thing that they eat: a one-stop, fully balanced meal plan for the Attini. Over time, the leaf-cutter ant genome has lost genes that were useful when the ant lineage lived independently from the fungus, genes that became useless in the context of the symbiotic relationship. For example, leaf-cutters no longer make the amino acid arginine. They now get this amino acid in sufficient amounts from the fungus. On the other side of the symbiosis, all fungal species cultivated by ants share one common ancestor, and all have evolved to produce specialized swollen hyphae called gongylidia that are easier for the ants to eat relative to the wild fungal ancestors.

The relationship between the ants and the fungi was not easy to figure out at first, but it was at least visible to the naked eye when Thomas Belt opened the underground nests. It was considered a prime example of symbiosis. For over a century, the fungal gardens of leaf-cutter ants were thought to represent a simple mutualistic partnership between two species. It was not until the end of the twentieth century that a modern microbiologist discovered that the macroscopic symbiosis between ants and fungi is only the first layer of a much deeper symbiotic network that extends far into the microbial world.

Fungal gardens are sensitive to infection by another fungus in the genus *Escovopsis*. It is a major threat: if *Escovopsis* comes into contact with the fungal

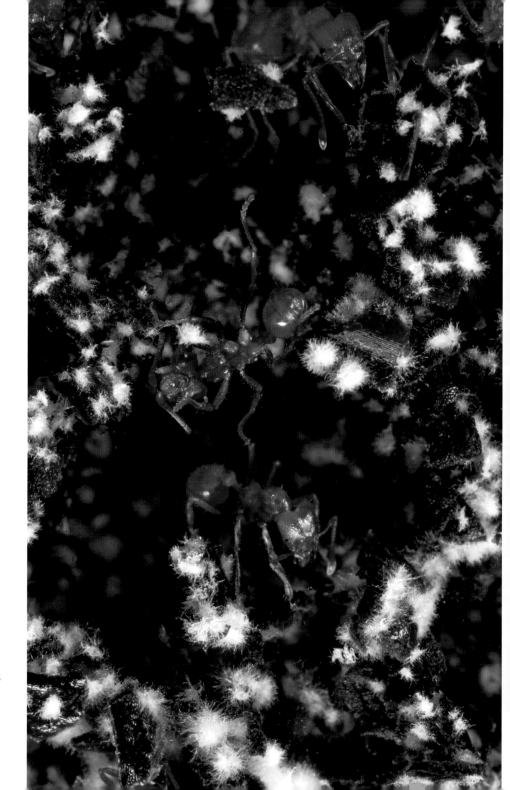

Leaf-cutter ants farm a crop of
fungus that grows on the leaves
they collect.

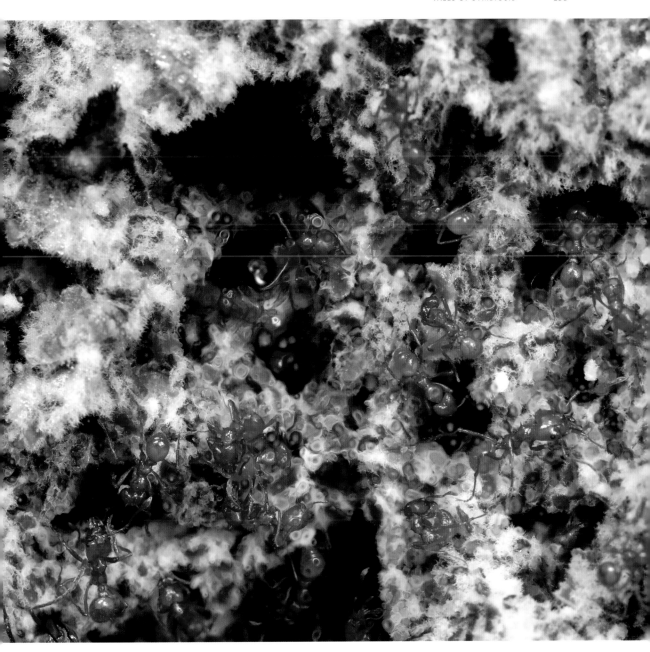

Another, even smaller, subtype of ants specializes in tending the fungal gardens.

crop, it can take over the entire garden. Scientists had observed that the leaf-cutter ants meticulously preen their gardens free of invading species. But this was not the full story. Ants have also evolved a symbiosis with some very special bacteria, members of the phylum *Actinobacteria,* which live right on the ants' exoskeletons. Actinobacteria are filamentous bacteria found all over the planet, particularly in soil environments. In particular, one group of actinobacteria, *Pseudonocardia,* lives on the ants, sometimes within specialized crypts of the exoskeleton and sometimes growing into visible fuzzy patches all over the ants. The ants do not make a chemical to suppress the growth of the pathogenic *Escovopsis,* but the actinobacteria do. The ant provides nutrients for the bacteria, and the bacteria, carried by the ants, produce antimicrobial compounds that keep the nest free of invading parasites.

The fifth species in the symbiotic network is a microbe that solves another problem for the ants. The leaves the ants collect provide plenty of carbon but not much nitrogen. Ants increase the nitrogen content of their crop in much the same way that humans do: they fertilize. Other bacteria, members of the species *Klebsiella,* make the fungal garden their home. These bacteria are capable of converting the inert nitrogen gas from the atmosphere into ammonia. In this form, nitrogen can be assimilated by the fungus, which is in turn eaten by the ant. The nitrogen is further recycled into the soil as dead ants are decomposed by yet other bacteria that eventually complete the nitrogen cycle by converting the nitrogen that is fixed in the soil into nitrogen gas. All told, this process is a major contributor to the biogeochemical cycling of nitrogen, not just within the fungal garden system but within the entire forest ecosystem.

Together, there are thus at least five species that interact to form a network of symbioses in every leaf-cutter garden. All relationships are mutually beneficial—the ant with the fungi, the actinobacteria with the ant, the ant with the nitrogen-fixing *Klebsiella*—except for the *Escovopsis* parasitic relationship with the fungal crop. Nonetheless, if we take a step back and look at the entire system, even *Escovopsis* is essential for the stability of the symbiotic network. If *Escovopsis* was not a constant threat lurking on the outskirts of the garden, the actinobacteria would likely lose the ability to produce antimicrobials over time—antimicrobials that may protect the colony from a variety of other parasites as well.

Now that we know that the leaf-cutter ants live in symbiosis with microbes that produce antimicrobial chemicals, it is a good time to stop and revisit our question about the ant stitches used by the Peruvian medicine woman. Did the ants she collected have enough actinobacteria on their heads to prevent other, more harmful bacterial or fungal species from infecting the wound? Was this natural antiseptic what made them effective as a therapy? One thing is clear: if there were actinobacteria on the ant heads, the medicine woman could not have known that they were there. It was a treatment that was passed down from generation to generation.

Soon, the entire world would realize the medicine-making abilities of microbes. Later in 1929, the Scottish biologist Alexander Fleming published the first report of the antibiotic penicillin. He had serendipitously discovered the compound when the mold *Penicillium notatum* contaminated one of his experiments and killed bacteria on a petri plate. But though it is a household name today, penicillin was not adapted as an effective drug for over a decade following its discovery. Fleming had trouble finding an easy way to purify the compound, and even when it was first used as a treatment in 1942, it was not effective against some types of bacterial infections, including tuberculosis, a leading cause of death at the time.

In the summer of 1943, an entirely different kind of underground activity was occurring. Albert Schatz was a twenty-three-year-old doctoral student at Rutgers University in New Jersey. He had just returned from duty at an Air Force hospital during World War II. During the war, he saw many people die from tuberculosis and became determined to find a drug to treat the disease. Schatz's advisor was a man named Selman Waksman, who ran a successful soil microbiology lab with the primary goal of discovering useful antibiotics from soil bacteria. Waksman liked Schatz's idea of working toward a cure for tuberculosis. However, he was afraid of going anywhere near the bacterium that causes the disease, *Mycobacterium tuberculosis*. So instead of working in Waksman's main labs, Schatz would work downstairs in the basement. Schatz accepted the conditions.

Schatz was hard at work in the basement laboratory from June until October of 1943. He would start with soil samples collected from various locations. Using petri plates containing nutrient-rich culture media, he isolated soil bacteria from these samples. In an experiment recorded in his notebook on

8/23/43 : Exp. 11. Antagonistic Actinomycetes (continued on page 49)

Control soil, Soils Nos. 2, 7A, 18A, leaf compost, straw compost, and stable manure plated out on egg albumin agar. Transfers made from colonies of actinomycetes selected at random (altho different representative groups were chosen — so far as this was possible by casual macroscopic observation). Some actinomycetes obtained from plates of swabs of chickens throats grown on egg albumin agar; these plates were obtained from Miss Doris Jones.

Material	Number of organism	Inhibition on nutrient agar		Inhibition on glucose-asparagin agar	
		Coli	Sub.	Coli	Sub.
I Control soil	C - 1	0 mm	0 mm	0 mm	0 mm
	2	1	2	1	4
	3	0	2	0	5
	4	0	7½	0	7½
	5	0	5	0	5
	6	0	0	0	5
	7	0	0	3	3
	8	0	0	5	5
	9	0	0	0	4
	10	0	0	0	2
	11	0	0	1	8

Albert Schatz's data from his notebook, "Experiment No. 11: Antagonistic Actinomycetes," 1943.

August 23, he noted that he was looking for one bacterial group specifically: the *Actinobacteria*. He worked to identify as many wild actinobacteria as he could, choosing them based on their multicellular, fuzzy-looking colonies, "so far as this was possible by casual macroscopic observation." He tested each species he found for an antagonistic interaction against known gram-negative and gram-positive bacteria: a phenomenon known as antibiosis. Each new isolate was systematically streaked in a line through the center of the plate. Any antibiotic molecules produced by the isolate would then flow out from the central streak and into the surrounding media. Meanwhile, the test species were streaked perpendicular to the central streak. If the isolate had no effect, the test species would grow all the way up to the central streak, producing no inhibition zone. If the isolate inhibited the test species, Schatz would simply measure the gap between the central species and the test species as a general measure of the inhibition strength.

Two isolates stood out to Schatz during that experiment. He marked organisms number 7 and 8 with arrows in pencil. These two isolates were cultivated from manured field soil near Schatz's lab and from the neck of a farm chicken. They were

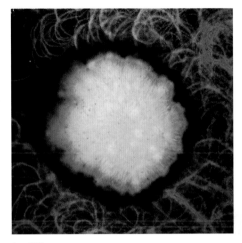

The phenomenon of antibiosis that Schatz was looking for: the bacterial colony in the center is inhibiting the bacteria growing around it. The species being inhibited by the central species is *Bacillus mycoides*.

effective against a broad spectrum of test bacteria, producing 3 to 5 millimeter inhibition zones. Both isolates happened to be species of actinobacteria from the genus *Streptomyces,* forming powdery colonies with water droplets on top that are filled with metabolites produced by the cells during growth. And both species killed *Mycobacterium tuberculosis.*

On October 19, 1943, a few months after he first observed antibiotic activity on petri plates, Albert Schatz knew that he had found a new antibiotic. He and Waksman named it streptomycin after the genus *Streptomyces.* The work was published in January, 1944. Tests in animals showed that streptomycin was not toxic, and they filed a patent later that year. It was the first treatment for tuberculosis and the next "wonder drug" after penicillin. Streptomycin and other drugs similar to it would go on to save many thousands of lives.

Neither ants nor humans invented these useful antimicrobial compounds. For human medicine, we isolate actinobacteria from natural environments, identify compounds with antimicrobial activity, and then mass produce that compound in industrial facilities. In contrast, ants carry around live actinobacteria that are constantly producing antimicrobials. Ants and humans use some of the same classes of antimicrobials from actinobacteria; humans take out the bacteria. By removing the microbe and dealing in purified chemicals we adapt antibiotics for many applications: as injections, pills, syrups, creams, soaps, you name it. For example, neomycin is another antibiotic discovered in the Waksman laboratory. It is one of the compounds commonly found in triple-antibiotic ointments that

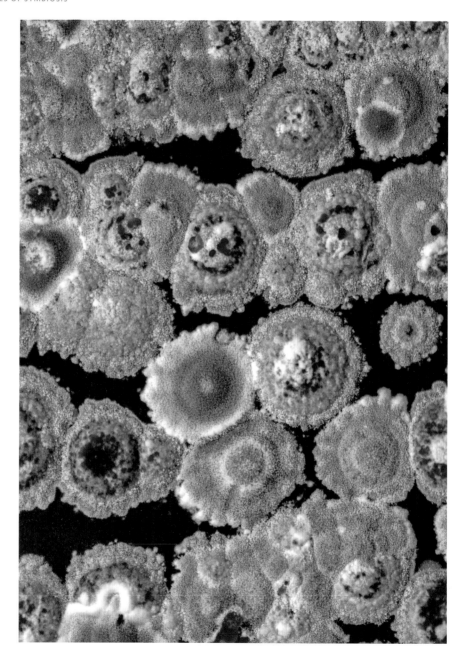

Colonies of *Streptomyces hygroscopicus.*

Streptomyces roseosporus seen with droplets of antibiotic.

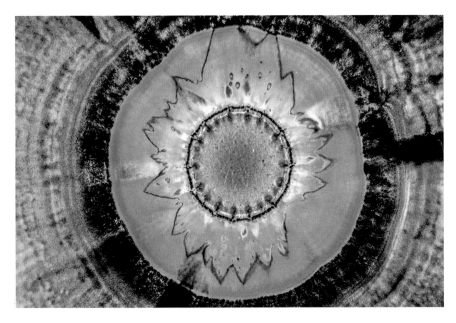

A mixed colony of *Streptomyces coelicolor* and *Amycolatopsis* sp. AA4.

we use to prevent infections after we scrape our knees. It was also isolated from actinobacteria—the same bacterial group that produces antimicrobials on leaf-cutter ant exoskeletons.

Removing the microbial middle man and isolating individual molecules for use as drugs may seem more precise and effective than the ant's way of using a living antibiotic-producing bacterium. And in some ways it is. The dozens of antibiotics that humans adapted from natural chemical products changed our daily lives and our societies. Infections that we now regard as easily cured were life threatening before the discovery and application of antibiotics.

But there is a serious disadvantage to administering antibiotic compounds in this way: the species they kill will almost always evolve resistance to those antibiotics, often very quickly. In nature bacteria produce antibiotics in low amounts and as complex mixtures of many different such compounds. Within their natural ecological context, the compounds we take and use as drugs may play an entirely different role. Antibiotics in nature often aren't found at concentrations high enough to kill bacteria and can act instead as nonlethal signals between different species. These molecules are part of living and changing systems. Both the species producing antibiotics and the species sensitive to those antibiotics are dynamic. Producers continuously evolve to make more potent compounds and receivers evolve to become immune or less affected by those chemicals. It has been described as an ongoing arms race.

This means that a mutation could occur in one *Escovopsis* cell in a leaf-cutter garden that makes it resistant to the antifungal molecules produced by the actinobacteria on the ant. The mutation may have modified the enzyme that the antifungal interferes with; it may have developed a more efficient efflux system for pumping toxic chemicals out of the cell; or it may be able to inactivate the antifungal molecule. In any of these scenarios, that one cell would suddenly have an advantage within the fungal garden ecosystem. It would be able to attack the crop fungus and could grow to overcome the entire garden, even in the presence of the ants. Nevertheless, there are most probably other ants in the same colony or in the neighboring colony harboring actinobacteria cells that have evolved the capacity to make a different antifungal that is effective against the new *Escovopsis* strain.

There is another kind of arms race at play between human antibiotic drugs and bacteria. A single drug compound administered by a physician is static. This is to say that when a bacterial strain develops resistance to that drug, it is up to human microbiologists and chemists to identify, isolate, and produce effective drug candidates. It is not an evolutionary arms race between human cells and bacterial cells causing the infection in a strict sense; it is an arms race between the resistant bacteria and cooperative human scientific efforts. Humans have also overused antibiotics. This was particularly true in the earliest days of drug discovery when antibiotics were applied unnecessarily as preventive medicine in livestock. Overuse accelerates the process of antibiotic resistance evolution simply because the more bacteria that are exposed to antibiotics, the greater the likelihood that one of those bacteria will become resistant. For this reason, there is now widespread resistance to all of the wonder drugs discovered by the Waksman soil microbiology group.

The evolution of antibiotic resistance comes down to the physical structure of the molecules involved at the nanoscale. Let's look at streptomycin. Though Schatz and Waksman discovered streptomycin in 1943, the way in which the molecule actually inhibits bacteria, known as its mode of action, was a mystery until the 1960s. That's when we learned that streptomycin and other related molecules, together called aminoglycosides, enter the cell and bind irreversibly to the ribosome, blocking protein synthesis. The British microbiologist Julian Davies found that streptomycin does this by binding within the active site of the ribosome. Normally, proteins are synthesized by transfer RNAs, selected by the ribosome with anticodons that each match the codons of messenger RNA. Streptomycin binds in the molecular space where this codon recognition step occurs, disrupting the cell's ability to correctly translate genetic information into functional proteins and leading to the death of the cell. Bacteria that develop resistance to streptomycin often have mutations in genes that encode the proteins that form the ribosomal recognition site. These mutated proteins thus lead to an altered structure of the ribosome, preventing the streptomycin molecule from interfering with protein synthesis and dodging the mode of action of the antibiotic.

An illustration aids the imagination as we picture in the mind's eye what such fleeting but profound interactions between molecules look like. Another antibiotic produced by the actinobacteria, actinomycin, kills microbial cells at the molecular level. Actinomycin works by intercalating into DNA—literally wedging into the DNA molecule. As it does so, it prevents the transcription of

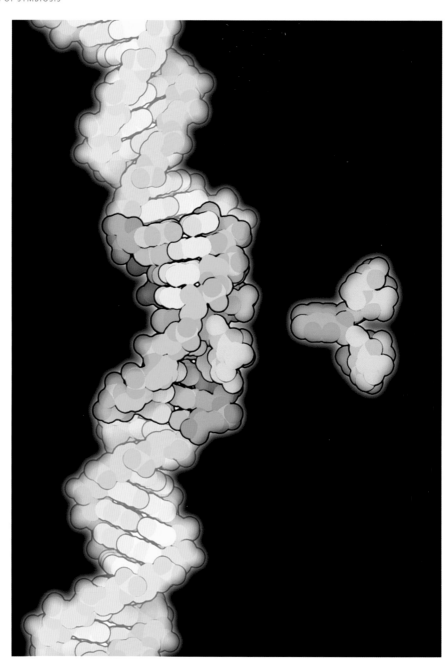

The molecule actinomycin, shown in blue and green in this accurate rendering, intercalates within DNA, blocking RNA transcription.

RNA by blocking the action of the enzyme RNA polymerase that elongates new RNA molecules. Without RNA, the cell cannot express any genes, and unless it has developed resistance to actinomycin, it will soon die. Unlike streptomycin, actinomycin cannot be used as an antibiotic in humans. Actinomycin is so effective at intercalating within DNA that it is cytotoxic, killing human cells along with any disease-causing microbes. Even so, derivatives of actinomycin have been developed as chemotherapeutic compounds to treat human cancers—yet another gift from the actinobacteria to humans.

Molecules are the interface of information exchange in the microbial world and the evolutionary currency through which species merge into a state of symbiosis. There is a virtually infinite pool of natural chemical products on Earth, most of them produced by microbes. To a degree that we do not yet know, the diversity of these natural products is a reflection of the unfathomable meshwork of interactions between species within biological systems. In order to establish a long-lasting symbiotic interaction, at least one species in a pair must contain or secrete at least one molecule that the other species uses to its advantage. But the advantages that those molecules bestow to each species can be very different. The advantage of the molecules could be simply that they can be consumed as life-sustaining nutrients, like the chemicals exchanged through the ant–*Leucocoprinus* symbiosis in the fungal gardens. An advantage could be that the molecule produced by the other species is an antimicrobial that keeps a parasite at bay, like the antifungals produced by the *Pseudonocardia* bacteria on the ant. So far we've seen these examples of symbioses on land, from microbes in the soil and in ant colonies. Let's look at several more symbioses in the ocean.

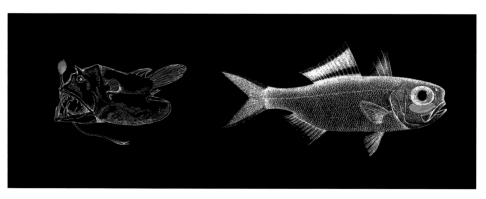

A monstrous anglerfish with dangling lure illuminated by bioluminescent bacteria.

Flashlight fish with cheek-like light organs filled with bioluminescent bacteria.

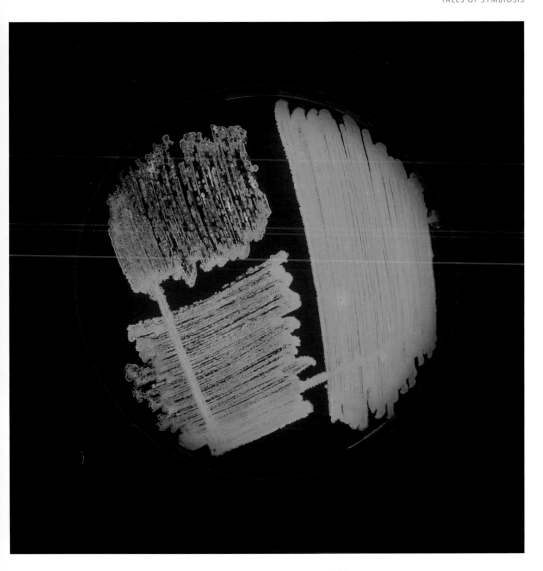

Bioluminescent bacteria *Vibrio fisheri* glow inside a petri plate when grown in the laboratory.

(Opposite)   The movements of a school of flashlight fish. A fifteen-second exposure captures the light released by *Photobacterium phosphoreum* in the light organ without seeing the bodies of the flashlight fish (*Anomalops katoptron*).

The bottom of the ocean is about as dark as dark can get. Yet the darkness of the abyss is often broken by the mysterious glow of many of its residents. The glow is caused by the chemical luminescence of biological molecules. Most fish and invertebrates in the deep sea glow in one way or another, and the light in many cases is produced not by the animals themselves but by symbiotic bacteria. There are dozens of species of flashlight fish and anglerfish, all with special light organs filled with bioluminescent bacteria. The anglerfish are among the natural monstrosities of Earth; with fangs sticking out from bizarre giant heads and a bacterial light organ that sprouts on a stalk in front of the eyes as a lure to bring their prey within striking distance. Flashlight fish have a different light organ; this one is under the eye and looks like a cheek. When a school of flashlight fish is photographed over a fifteen-second period, each individual leaves a streak of light showing where it has been.

The cheek pocket of a flashlight fish is filled with millions of *Photobacterium phosphoreum* bacteria that produce light through a chemical reaction. When the bacteria are cultured in the laboratory, the light produced by this reaction causes the entire petri plate to glow blue-green. *Photobacterium* cells live in the fish cheek or culture flask at high density where they use quorum sensing to coordinate expression of *lux* genes. The *lux* genes encode proteins that work together to produce light, and they are arranged into a cluster within the genome that is called an operon. The genes *luxA* and *luxB* encode two polypeptides that come together to form the luciferase enzyme. The LuxA domain (blue) contains the catalytic core of the enzyme, but it only works effectively when the green LuxB domain stabilizes the entire structure. And we know it is working effectively when it emits light upon oxidizing its substrate, decanal, into decanoic acid. Flashlight fish gain a variety of advantages from this bacterial bioluminescence by controlling light output with membranes that open and close over the light organ like an eyelid. One pattern of flashing light may be used to attract zooplankton as prey, another to confuse attacking predators, and another to communicate with other flashlight fish.

Similar light organs filled with a closely related bacterium, *Vibrio fischeri*, are found in the Hawaiian bobtail squid, *Euprymna scolopes*. Here we have examples of mutually beneficial symbioses that are again based on chemical exchange. The symbiotic bacteria grow using metabolite molecules produced by the host animals, and the animals benefit from light emitted by bacterial luciferase.

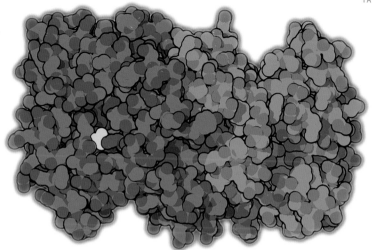

The protein bacterial luciferase produces bioluminescence.

In this case, light emitted at night, when the squid is most active, prevents it from appearing as a dark shadow against the moon when observed from below by a predator. In the squid-vibrio system, the symbiosis is so tight that the tissue of the squid lining the light organ is actually remodeled in response to the bacteria in order to maintain a stable daily cycle and a stable light organ for use at night. Humans have benefited from this symbiosis by copying *lux* genes from oceanic bacteria and using light production by luciferase as a biotechnology to screen for toxins and visualize tumors in experimental animals.

Along with bioluminescence in the deep sea, marine animals that live closer to the surface have formed symbioses with photosynthetic microbes. Microbes called zooxanthellae are found in corals, anemones, jellyfish, giant clams, and sponges, among other invertebrates. Zooxanthellae act like the bacteria that fix nitrogen gas and cycle nitrogen into leaf-cutter gardens, only they are not bacteria, they are eukaryotic microbes called dinoflagellates. And instead of fixing nitrogen gas, they use photosynthesis to fix carbon dioxide, bringing carbon into coral ecosystems.

The most common of the zooxanthellae are golden brown cells from the genus *Symbiodinium*. There are millions of corals in the Great Barrier Reef off the coast of Australia, and most of those corals contain many millions of invisible *Symbiodinium* cells embedded within their internal tissues, all lined up to maximize exposure to light. In just one coral species from the genus *Pocillopora*, found in the Great Barrier Reef and other parts of the Pacific Ocean, we see with light microscopy a group of golden *Symbiodinium* cells along with some common coral cells called nematocysts. The translucent nematocyst cells contain a curled-up stinging organ. In spite of their passive appearance, corals are predators. The tentacles of each coral polyp extend to trap zooplankton during feeding, injecting toxins from nematocysts into tiny zooplankton and dragging them, paralyzed, into their (comparatively) giant mouths. The epicenter of symbiosis between the coral and the *Symbiodinium* cells is the symbiosome, a membrane-lined pocket produced by the coral. The symbiosome space surrounding each *Symbiodinium* cell is a microenvironment controlled by the coral. In this way, the coral can manipulate the physiology of their endosymbionts, maintaining them in conditions that are ideal for photosynthesis.

This is one of the most intense forms of symbiosis in nature—a nested endosymbiosis. The *Symbiodinium* cells themselves did not invent photosynthesis. In their own evolutionary past as free-living dinoflagellates, another photosynthetic eukaryotic microbe was engulfed. And that eukaryotic microbe had itself in its evolutionary past engulfed a photosynthetic alga, which had itself previously ingested a photosynthetic cyanobacterium: the cyanobacterium that became the chloroplast. It is the bacterial ancestor of that chloroplast, inside the alga, inside the zooxanthellae, inside the coral, that first evolved the photosynthesis that feeds the coral today. Looking at the coral as a whole, what we have is a merging of no fewer than four organisms over time. Small or not, zooxanthellae are a keystone species: they have a remarkable impact on the environments they live within, relative to other species. They are the foundation of all coral reef ecosystems on Earth. These reefs are the rainforests of the ocean, home to thousands of animal species. When a coral reef is disrupted and bleaches, turning from golden brown to white, what has actually happened at the microscopic level is that the corals have lost their zooxantheallae.

We can even imagine corals and other organisms that live together with symbiotic microbes as metaorganisms, or holobionts. These two concepts are both ways to think about corals, bobtail squid, and perhaps also humans and their associated microbes as cohesive functional or ecological units. The idea was first proposed by Lynn Margulis in 1991. Within this conceptual framework, the human genome encodes from our own chromosomes some tens of thousands of genes. Our collective hologenome, adding all the microbial genomes within our bodies, contains millions of genes.

The 2,300-kilometer Great Barrier Reef is among the largest living ecosystems, providing a home for over 1,500 species of fish. It is one of the living systems on Earth that can be easily seen from space.

A coral from the genus *Pocillopora* with green polyps partially retracted.

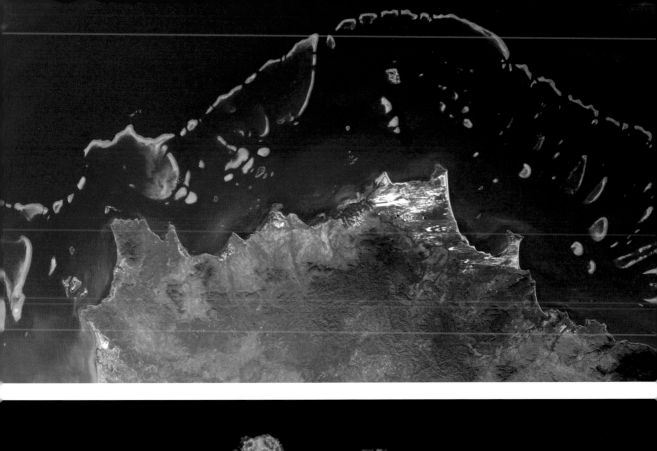

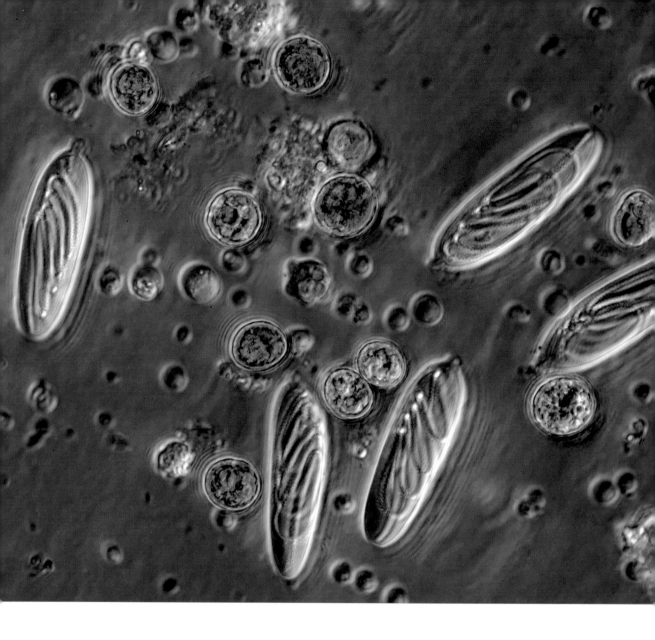

Golden zooxanthellae dinoflagellate cells alongside coral nematocysts.

Symbiosis is a spectrum from the most intimate forms of mutualism to the deadliest cases of parasitism, like the bacterium *Mycobacterium tuberculosis* in humans, and everything in between. Every long-lasting relationship between two or more organisms falls somewhere along this spectrum, and those positions are not fixed. An organism's relationship with another organism can be very

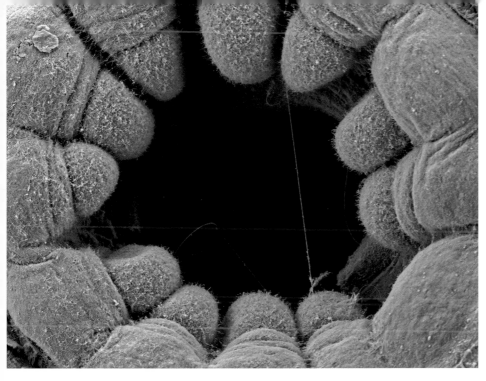

The mouth of a *Pocillopora* polyp. Each retracted tentacle is lined with stinging nematocysts.

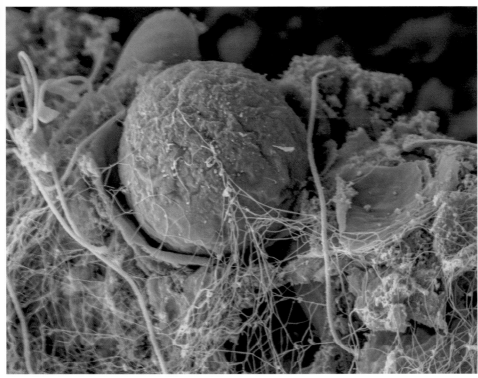

A zooxanthella within a symbiosome membrane inside fractured coral tissue (endoderm).

stable, like the mutually dependent ants and their fungal crops, the flashlight fish with their bioluminescent bacteria, or the corals with their photosynthetic zooxanthellae. Yet symbioses can also change rapidly. Right in our own nasal cavities, most people harbor two species of bacteria, *Streptococcus pneumoniae*, which can sometimes cause harm in humans, and *Corynebacterium accolens*, which never harms humans. The *Corynebacterium* bacteria produce fatty acid molecules that inhibit the growth of *Streptococcus pneumoniae*. So long as *Corynebacterium* is present within the nasal microbiome, *Streptococcus pneumoniae* is commensal; it exists there and causes no harm. But if *Corynebacterium* is removed, *Streptococcus* shifts from commensal to parasitic.

The relationship between any two species may change between mutualism and parasitism as easily as human relationships can fall out of balance. We can think of this kind of change by returning to the story of Albert Schatz and Selman Waksman, the co-discoverers of streptomycin. Schatz and Waksman's relationship as advisor and student began well. However, in the following years, Schatz felt that Waksman minimized his role in the discovery. At first, Waksman failed to mention his involvement as he accepted awards. Waksman was on the cover of *Time* magazine and appeared on postage stamps. Then, as royalties began to arrive on the sale of streptomycin, Schatz realized that Waksman had left him out of the contract, initiating a contemptuous legal battle. Finally, in 1952, the Nobel Prize was awarded for the discovery of streptomycin to a single scientist: Selman Waksman. The Nobel committee had rewarded Waksman for his pioneering methods that led to the discovery, but not credited Schatz for the work behind the discovery itself. Ultimately, the story had a happy resolution: in the 1990s, there was renewed interest in recognizing Schatz for his discovery of streptomycin. By then in his seventies, Schatz was presented with the Rutgers University Medal in 1994. Although Schatz and Waksman never personally reconciled, one fact remains true for humanity: for every relationship that turns sour, another new and synergistic relationship is formed.

Human society is built on a web of ever-changing relationships. Like ants, humans are among the most social organisms on Earth. E. O. Wilson, who founded the field of sociobiology, considers both ants and humans as eusocial species, meaning that we engage in the most advanced form of social organization seen in nature. Within our societies, relationships are dynamic. We care for each other like no other species, and we compete with each other and within social groups like no other species. For humans, bonds and breakups are formed and fought through the mind, through words and thoughts that manifest in action. In eusocial insects, a change of chemical cues can mark an old or invading ant for immediate intertribal assassination.

Societies of humans and ants act like superorganisms. Advanced systems of cooperation have enabled us to use other species to our advantage, whether it is medicine passed down through the culture of indigenous peoples, the mushroom gardens of ants, or the human farms that cover so much of Earth today. Together with our respective symbiotic microbes, ants and humans are the two most successful animals on Earth. Yet there is a critical difference in the trajectories of our success. Ants have a mutualistic effect within the ecosystem. Plants near leaf-cutter nests are more abundant, grow faster, and have more robust root systems than plants in areas without nests. Humans, however, often have a net negative effect on natural ecosystems. Can we bring back into balance our runaway technological achievement and population growth with our impact on the biodiversity of Earth?

We can, so long as we appreciate natural ecosystems as the setting of our own evolution, from the first microbes all the way to *Homo sapiens*. We can if we see the biosphere for its natural beauty, for the symbiotic approaches evolved by species living in rainforest canopies, in coral reefs, and in our backyard. We can continue to collect ants for natural stitches, actinobacteria for the antibiotics they produce, and many other species in the future as long as we conserve biodiversity and appreciate all forms of life, large and small. We must let our knowledge of biology guide our path as a species to coexist within the nurturing ecosystems of Earth.

A commemorative stamp issued by Gambia in 1989 celebrated the discovery of streptomycin.

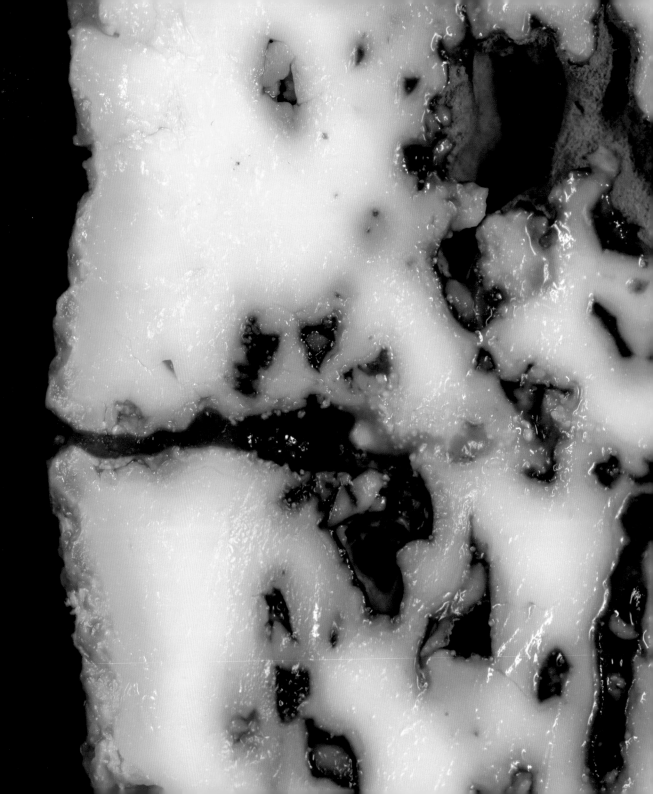

6

On the Kitchen Counter

THE STUDENT SPENT ALL DAY on her studies of microbes and food. Exhausted and hungry, she couldn't help but contemplate the food she readied for her evening meal. As she gazed at the blue cheese, the wine, the sautéed mushrooms and sliced Italian salami, and the chocolate-covered espresso beans for dessert, her mind began to wander. Lost in a daydream, she thought, "What if I could shrink myself so small that I could climb up from the surface of this kitchen countertop and onto that mountain of blue cheese?" She imagined her goal on this fantasy trek: she would dig into the rind and through the cheese like an archaeologist on an expedition to an unexplored world. After tunneling deep into the cheese, she would discover treasure—veins of blue mold.

She was searching for caverns filled with *Penicillium roqueforti,* a species of cheese mold that forms fuzzy, blue-green colonies. First she would need to climb across the thick microbial forest of the rind to reach the entry point. Then she had to spelunk into an abandoned tunnel and blast through centimeters of cheese curd. All the while, she needed to watch out for microscopic animals that live in the cheese. And all she carried was a map, a torch, a satchel to collect the *Penicillium,* and a bundle of miner's dynamite.

Inside the rind, she pushed through fungal mycelia that crisscrossed above her like thick forest vines. These are the multicellular filaments of the white molds that grow on the rind. In between the mycelia, there were white, yellow, and orange colonies of bacteria and yeasts. In all, she saw dozens of different species of bacteria and fungi growing below the mycelia. It is a view of a blue cheese rind that anyone can see with a scanning electron microscope. This is the microbial community that grew on the rind as the cheese aged in a damp, cool cave for several months. Some of the microbial species present, like the white mold that formed the tall mycelia, were intentionally added by the cheese maker. Many other species arrived on the cheese by their own devices from within the cheese-cave environment.

She looked down at her map and realized she had reached the entry point. There was the entrance to the old tunnel that led to the interior of the cheese.

Veins of blue mold within the tunnels and pockets inside blue cheese.

The rind of a blue cheese.

(Opposite, top) The microbial community on a blue cheese rind seen with electron microscopy.

(Opposite, bottom) Dozens of bacterial species and fungal species grow on the rind.

It was one of the locations where the cheese makers poked the rind with a long metal needle to let oxygen into the maturing curd. But there would still be places where the cheese had collapsed, and dynamite would be needed to clear the debris. Into the blue cheese she went.

It did not take long for the light from the microbial forest above to fade away. A few millimeters in, she was at the base of the rind, where enzymes secreted by the microbes on the surface break down nutrients in the cheese. It was the part of the cheese where metabolic by-products from the microbes growing on the rind accumulate. And it is those by-products that give different types of aged cheeses their unique flavors. As she traveled farther into the tunnel, the cheese turned from a darker yellow to almost pure white. She was now in the interior curd of the cheese, too far from the rind for the microbes growing there to have

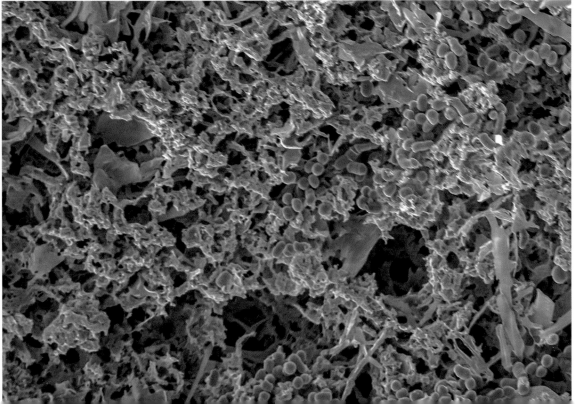

Wheels of blue cheese aging in a cheese cave.

much influence. A different group of microbes held sway here, those that influenced the first stage of cheese making.

She lit her torch and looked at the curd along the wall of the tunnel. All of the cheese curd she saw began with milk that had been transformed through a process of microbial fermentation. The milk was first separated into liquid whey and solid curds through the activity of lactic acid bacteria like *Streptococcus thermophilus* and *Lactobacillus* species. These bacteria were added to the milk as a starter culture. As the bacteria grew, they converted lactose, the milk sugar, into lactic acid. As more lactic acid accumulated, the milk became acidic, and casein, the most abundant milk protein, coagulated. At this time the curd was collected

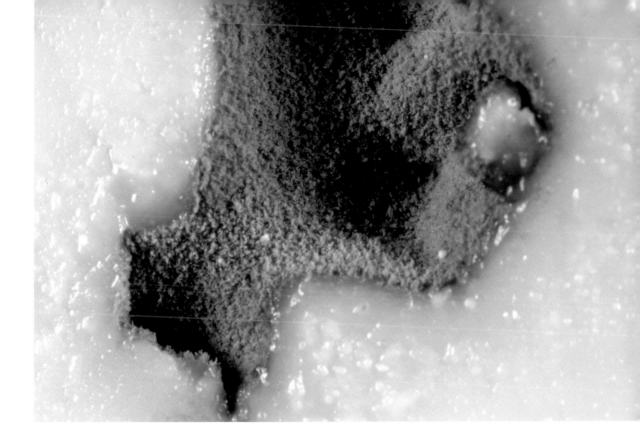

and squeezed into its familiar round shape. From this point, some cheese curds are dried, some are cooked, and some are salted. More microbes are introduced to some cheeses as adjunct starter cultures to add extra flavors or properties. For example, the bacterium *Propionibacterium freudenreichii* creates the holes in Swiss cheese by producing carbon dioxide gas. Some cheeses are aged, and some, like mozzarella, are consumed without any aging. Blue cheeses are not cooked prior to aging. That meant the white curd she was exploring in the interior still contained thousands of live lactic acid bacteria. The lactic acid bacteria can live anywhere within the cheese because they can grow anaerobically, without any oxygen.

The explorer saw a hint of blue behind a wall of white cheese. At last! The area was glowing blue-green with *Penicillium roqueforti*. There was only one problem: it was blocked off by a collapsed cheese curd. She grabbed her first stick of dynamite, placed it carefully at the center of the curd, lit the fuse, and ran back as fast as she could. The explosion blasted cheese everywhere, and as

A cavern of *Penicillium roqueforti* within blue cheese.

Cheese mites live on just about any kind of aged cheese. Mimolette, a French cheese, derives its distinct character from the large pockets in the rind carved by cheese mites.

the smoke cleared, she saw that the tunnel was passable. She walked back down the tunnel, excited to enter the cavern and collect the mold. But just as she was about to enter, a giant cheese mite emerged.

There it was, *Tyrophagus,* crawling right in the center of the tunnel. From the same class of arthropods that includes the spiders and scorpions, these mites are the fauna of the cheese microbiome. It meant no harm to her, however, because these mites are not carnivores. Cheese mites are comparable to cows: grazers that eat the fungi in the cheese instead of eating grass. Mites help the fungi by dispersing fungal spores and by digging through the cheese, creating new surfaces for the fungi to grow on. In turn, the mites have a continuous food supply. And most important (from our perspective at least), mites contribute to the flavor of the cheese by producing a lemony tasting compound. But when you are ten micrometers tall, mites are monsters. The mite could run over her without knowing it or maybe just slurp her up by accident. Either way, it wasn't a risk she was willing to take. She grabbed another stick of dynamite, lit the fuse, and threw it as hard as she could, sending it directly into the mouth of the arachnid beast. But just as she thought she was saved, the fuse went out. The mite was

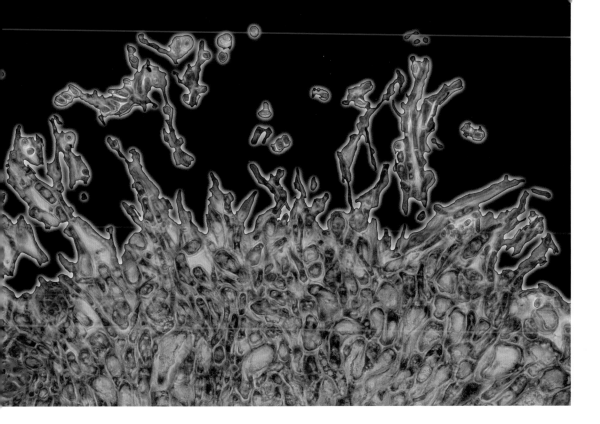

about to trample her as she lit the last stick of dynamite and aimed it at one of its legs. It exploded just in time, scaring the mite away and out of sight down the tunnel.

As she stood at the edge of the cavern, she held up her lantern and saw the blue *Penicillium roqueforti*. It grew in threadlike hyphae like the mold on the surface of the cheese, with stalks called conidiophores branching in many directions. Each conidiophore was the shape of a bouquet, filled with chains of reproductive spores or conidia. These round conidia become apparent when looking at *Penicillium* conidiophores using a light microscope and better still when they are stained pink for greater contrast. The species is a microaerophile: it needs less oxygen than the species on the surface but more than the anaerobes inside of the cheese curds, using whatever oxygen filters down through the tunnel. Here was the treasure—the source and explanation of how blue cheese came to be.

The same species of *Penicillium roqueforti* has been used by cheesemakers for over a thousand years. Legends of the first blue cheeses trace back to natural *Penicillium roqueforti* molds that still grow in caves at Roquefort-sur-Soulzon in southern France, where the original blue cheese is made. But the strain the explorer collected in the tunnel of this modern blue cheese is different. Like the lactic acid bacteria starter cultures used to produce the cheese curd, this *Penicillium roqueforti* was added to the cheese intentionally by the cheese makers.

Strains of *Penicillium roqueforti* found on most commercial blue cheeses today have been domesticated. Generations of cheese makers have selected and propagated strains of mold with particular desirable traits over time. They have chosen molds from cheeses that may have developed a new, appetizing aroma, or a rich, gooey center. Or maybe they preferred an aesthetic quality, like color. On every blue cheese rind, there will be a smaller number of *Penicillium* cells that have developed mutations in genes affecting the natural color of the mold. By sampling the cheese and growing the *Penicillium* species in the laboratory on solid growth media, we can recover these mutants. All of the *Penicillium* colonies on the resulting petri plate are the same species, but one emerges as brilliant white among the ancestral green colonies. This is evolution in plain sight, from green to white. Now imagine that one of these white strains developed into a large colony naturally on a rind within a cheese cave. If the cheese maker liked the white color more than blue-green and chose it for the next batch and the next, this is all it would take to begin domesticating a new strain of *Penicillium* that might be found in cheeses centuries later.

When the genomes of domesticated and wild strains of *Penicillium roqueforti* are compared today, some areas of the genome stand out. Several sections of the domesticated *Penicillium roqueforti* genome contain genes from *Penicillium camemberti,* a closely related mold species that gives Camembert cheese its unique qualities. The presence of these genes indicates that there was a precise moment in the past when the exchange of genes occurred: a moment known as a horizontal gene transfer event. The *Penicillium* genes may have been swapped when the conidiophore stalks of the two mold species fused and transferred nuclei from one cell to another during conidial anastomosis, a particular mechanism of horizontal transfer common in filamentous fungi. Some of the transferred genes are important in cheese making, including a lactose permease gene encoding an enzyme that enhances the ability of *Penicillium* species to live within the milky cheese curd.

(Opposite) A white colony appears among the blue cheese microbes cultivated in the laboratory by Benjamin Wolfe. The large mold colonies are a close relative of *Penicillium camemberti*. The white colony is a mutant that evolved from the green ancestor. The smaller, smooth colonies are the yeast *Debaryomyces*.

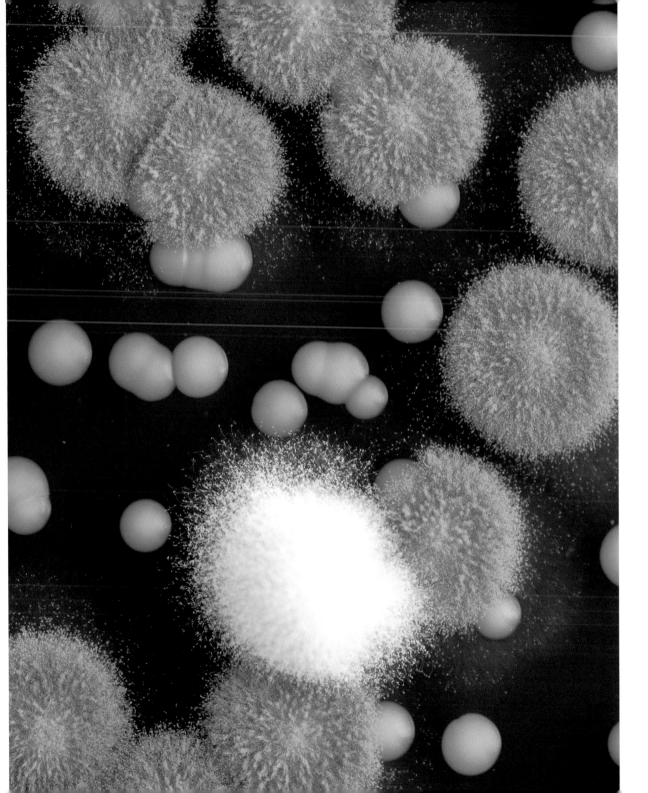

The student explorer broke off a piece of the blue mold and put it in her satchel, ready to return from her fantasy adventure. As she imagined herself walking out of the tunnel with the artifact, she suddenly snapped out of the day-dream. She was back in her kitchen, contemplating an ounce or two of cheese. She poured herself a glass of wine, wondering as she took a sip what stories might be contained within her glass. Once more she was off, daydreaming about the microbes that transformed her glass of grape juice into a tumbler of merlot.

Both beer and wine are produced by another fungal species in the same phylum as *Penicillium* species, the yeast *Saccharomyces cerevisiae*. This is common baker's yeast, the same tiny round cells that convert sugars into carbon dioxide when added to flour dough, causing the bread to rise. *Saccharomyces cerevisiae* is also called brewer's yeast because the metabolic reaction that produces carbon dioxide also produces the ethanol in beer and wine, the ethanol whose pleasant effects are experienced by individuals who consume those beverages in moderation. Humans have been drinking alcohol and enjoying the euphoria it can induce since the Stone Age. Organic chemical residues from a fermented beverage made from rice, honey, and fruit have been discovered in pottery made in a Chinese village 9,000 years ago. Other residues found in ancient Egyptian jars indicate that *Saccharomyces cerevisiae* was used to make wine there at least 5,000 years ago.

Cells of the yeast *Saccharomyces cerevisiae.*

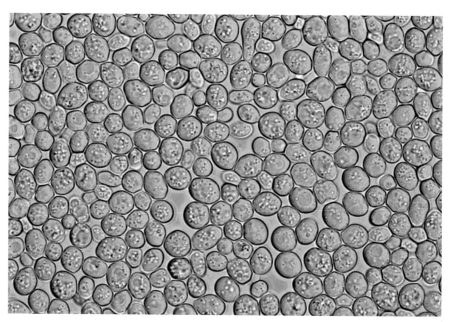

As different peoples used *Saccharomyces cerevisiae* to produce beers and wines across the planet, they began to use leftover cultures from previous fermentations as starter cultures for subsequent fermentations, finding that this practice of back slopping would lead to faster fermentations and more diversity in flavor. This custom kick-started the domestication of *Saccharomyces cerevisiae* cells in much the same way as *Penicillium* species were domesticated in blue cheeses and Camembert cheeses. Despite the similarity in process, however, *Penicillium roqueforti* and *Penicillium camemberti* are both more or less equally domesticated species, whereas *S. cerevisiae* strains in wine and beer are not.

Beer is produced in large vats, where yeast cells live in a stable, nutrient-rich medium. The strains of *Saccharomyces cerevisiae* experience nothing other than this environment and grow continuously throughout the year without any off-season or change in the industrial niche in which they live. Every time a brewer makes a new batch of beer, yeast that settles to the bottom of the prior fermentation is added. In doing so, the fittest yeasts for the man-made growth conditions are constantly being selected. In contrast, in traditional wine production the strains of yeast are involved in alcoholic fermentations only after the harvest of the grapes, once a year. In between harvests, these yeasts must retain their ability to survive in natural environments. Each time that grapes are harvested, new yeasts are introduced from the vineyard.

For this reason, when the genomes of more than one hundred strains of *Saccharomyces cerevisiae* were compared, those found in grape fermentations were far more wild than those found in beer fermentations. The beer yeasts were efficient fermenters of the common ingredients used to make beer through mutations in genes responsible for the degradation of sugars found in grains, predominantly maltose. Beer yeasts lost the ability to produce several chemicals responsible for off-flavors in the beer. Many beer strains have even lost their ability to reproduce sexually. The ability to form spores that survive long periods of stress has no advantage within continuous man-made cultures, causing the genes for this process to have been lost over time as well. This means that the yeasts used in making beer have become isolated from wild yeast strains and are fully dependent on beer making for their own reproduction. Heavily domesticated, the yeasts rely entirely on humans for their reproductive success. In contrast, wine-producing strains have picked up some domesticated traits over the centuries, though not nearly as many as beer strains. They can still mate with wild yeasts and reproduce successfully without humans. This pattern may be changing, however, as wine producers seek greater control over their finished product. In recent decades wine producers have begun to kill the yeasts resident on the harvested grapes and introduce yeast strains that have been grown in laboratories before fermentation. We'll have to wait and see how the domestication of those winemaking yeasts proceeds in the future.

Our daydreamer in the kitchen continued to prepare dinner, chopping white button mushrooms. She sautéed them in butter with crushed garlic and a pinch of herbed salt. She knew that her mushrooms are the fruiting bodies of the fungal species *Agaricus bisporus*. These mushrooms that we see and eat are but one stage within a life cycle that extends into the microbial world.

Every new patch of white button mushrooms begins with microscopic basidiospores that germinate and develop into still-microscopic hyphae. Hyphae begin in a haploid state, possessing only half of the chromosomes needed to make mushrooms. As hyphae form threadlike networks within the substrate, they contact and fuse with other genetically compatible mating types. After fusion the hypha cells have two nuclei (one from each parent hypha) and a complete set of chromosomes. Fused hyphae become thicker mycelia that grow inside of logs or on artificial growth substrates in mushroom farms. Mycelia soon begin forming fruiting bodies that grow rapidly into the buds that we know as mushrooms. Mushrooms develop further into dark gills that form beneath the cap. Each finely arranged gill structure is lined with smaller basidia structures: it is these basidia that produce and release more of the haploid basidiospores, completing the life cycle.

Agaricus bisporus is another microbial species that has been domesticated by humans. In the late 1920s, one mushroom grower in Pennsylvania came upon a cluster of mushrooms on his farm that had white caps rather than brown caps. Because he felt white mushroom caps would be perceived as more attractive in grocery stores, he chose this cluster of white variants to propagate in all of his gardens. This color change was caused by a single alteration in the mushroom genome. That strain from the 1920s is now the main commercial variety of button mushrooms, produced worldwide by the millions of tons. Oyster mushrooms and shiitake mushrooms have also been domesticated.

Our tastes so often change over time. The highly prized mushrooms are no longer those common white buttons. Now there are several brown varieties of *Agaricus bisporus* marketed as cremini and baby bella mushrooms. If these

Agaricus bisporus hyphae form networks of mycelia.

same strains are left growing until their caps reach full developmental maturity, they are called portobellos. We know portobellos as the large and delicious dark brown caps with fully developed gills underneath. The gills are packed with microscopic spores that complete the organism's life cycle upon release into the air. All of these varieties of *Agaricus bisporus* arise from minor genetic differences in the same fungal species.

The food microbes in the student's dinner on the kitchen counter have all undergone processes of domestication by human artificial selection similar to the processes involved in the domestication of animals and plants. Look, for example, at the amazing variety in dogs, or squashes. Still, there is one big difference. Animals and plants have been intentionally domesticated over longer time periods. The common ancestor of dogs and wild canines lived some 40,000 years ago. One species of wild squash, the predecessor of all of our winter and summer squashes, was domesticated over 10,000 years ago. Many of the genetic changes that led to the domestication of food microbes have occurred in only the past few hundred years. Yet in that shorter time period they have evolved rapidly. So, domesticated microbes have been on an evolutionary fast track.

The first reason that microbes are able to adopt genetic changes so quickly is simply that many of them can reproduce rapidly. Evolution acts in units of

Mycelia bud into fruiting bodies.

generations. While a human generation may last twenty-five years, a microbial generation can last twenty-five minutes or even less. Some domesticated beer yeasts, for example, have likely grown more or less uninterrupted for over 75,000 generations. Rapid growth rates and short generation times lead to a large number of individual microbial cells. Every cell within populations of many billions of cells is its own evolutionary experiment. If a mutation that increases fitness occurs in just one cell in that population, that mutation can spread very rapidly, leading to what is known as a selective sweep. That variant gene can spread as the cells increase their abundance in the population by out-competing other cells for resources or by surviving an environmental challenge.

Genetic traits can also spread quickly among microbial populations through horizontal gene transfer. Microbes on a cheese rind can exchange genetic information with each other in several ways. We investigated mycelial fusion in *Penicillium roqueforti,* but there are other mechanisms for gene transfer in microbial species. One major mechanism is the uptake and incorporation into the genome of free DNA from the environment, a process called natural transformation. A second mechanism is known as conjugation, which depends on narrow

Fruiting bodies grow rapidly into larger buds.

Buds at this developmental stage
are collected as white button
mushrooms.

This mature mushroom is now considered a portobello because it is large, brown, and has formed gills lined with reproductive spores.

projections called pili on the surface of cells. Pili allow cells to make contacts that facilitate the movement of DNA from one cell to the other. A third major mechanism of horizontal gene transfer is transduction, when a virus infects a cell and then the virus progeny pass along pieces of that cell's genome to the next cells that they infect.

Through this process of domestication, microbes have aided human societies just as domesticated plants and animals have. Much of the global food production that supports over seven billion humans on Earth depends upon microbes. Microbes have been on and in our foods all along, from our days as hunters and gatherers, to when we began to farm and keep livestock, to the present day. Thousands of years before any person had any notion of the microbial world, microbes were hard at work producing human food and drink.

A kombucha biofilm or SCOBY develops above fermenting black tea. The SCOBY develops where the air and liquid meet during a fermentation process lasting one week or more.

Cheese, wine, and button mushrooms are only a few examples of microbial foods. Kombucha is a fermented tea produced by a thick floating biofilm nick-named SCOBY, which is an acronym for "symbiotic community of bacteria and yeast." If we look closely at a jug of kombucha tea, zooming in to see the floating SCOBY biofilm and then even closer, observing discrete cells using a scanning electron microscope, we can see the individual bacteria, the yeast, and the gluey microbial cellulose fibers that bind the biofilm together. Not all kombucha biofilms look exactly the same. As complicated ecosystems, SCOBYs vary from kombucha maker to kombucha maker. Some microbial groups are common across many SCOBYs, such as yeast from the genus *Zygosaccharomyces* and

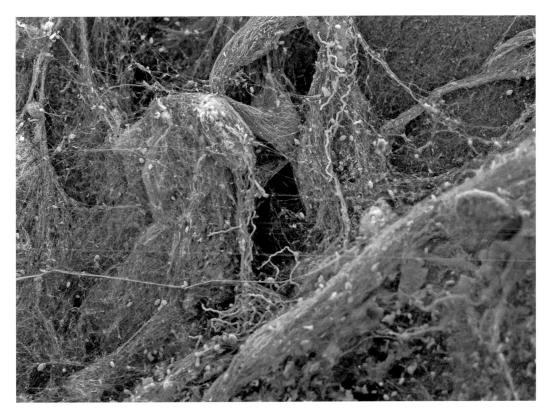

Cellulose fibers that bind the biofilm together seen with electron microscopy.

the bacteria that produce the cellulose fibers (often from the genus *Gluconace-tobacter*). Less common species like filamentous fungi can also find their way into the biofilm, imparting unique flavors and aromas. Beyond flavoring the tea, SCOBY microbes help preserve the beverage by increasing its acidity. There are many more popular fermented foods. For example, lactic acid bacteria inside the interior of cheese are critical not only for the production of cheese curds, but also for the production of buttermilk, sour cream, kefir, and yogurt. Kimchi and sauerkraut are fermented forms of cabbage, produced through the activity of microbial species naturally found on cabbage leaves. Microbes are also needed for the curing of meat products like salami.

Yeast, bacteria, and filamentous mold seen inside the biofilm at the microscale. Based on metagenomic studies of kombucha biofilms, the most abundant, short, rod-shaped bacterial cells seen in the image are likely from the genus *Glucon-acetobacter*, which produces the microbial cellulose fibers. The yeast cells are likely from the genus *Zygosaccharomyces*.

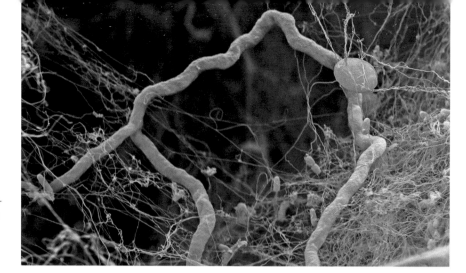

Budding yeast cells enmeshed within crisscrossed microbial cellulose fibers. In this image, the reproductive buds can be clearly distinguished on the yeast cells.

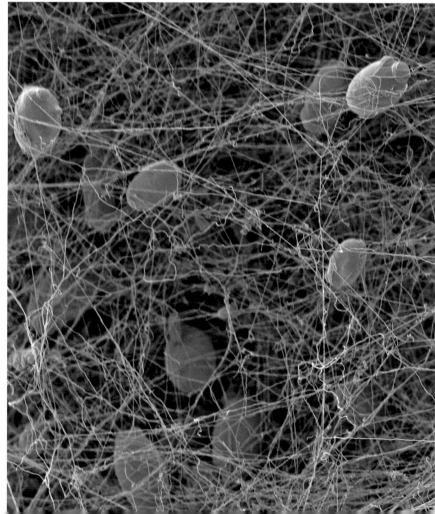

(Opposite) Microbes cultured from the cabbage leaves used to make fermented kimchi. A clear zone surrounds the two pink-orange colonies in the center, indicating that this species has the ability to break down casein within the growth medium.

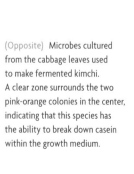

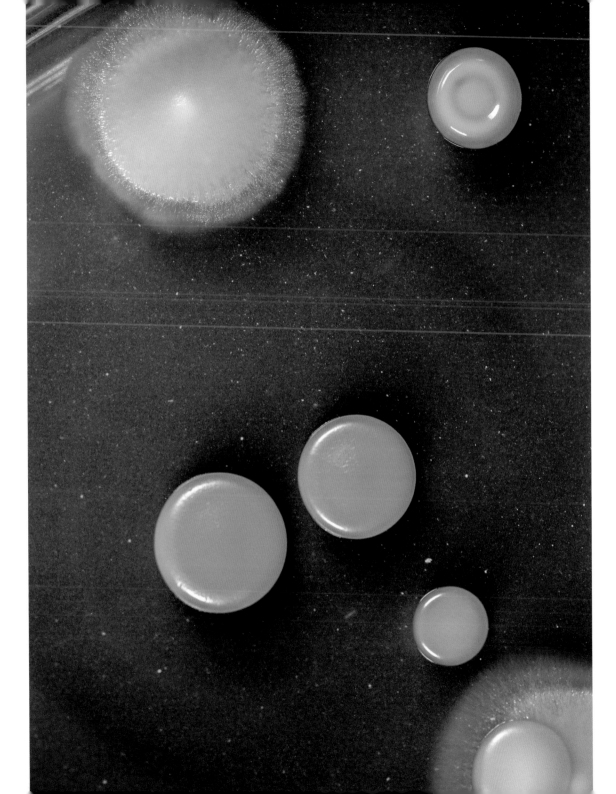

Aspergillus oryzae forms filamentous structures.

(Opposite) *Aspergillus oryzae* growing on rice kernels.

Another mold, *Aspergillus oryzae,* has been used for thousands of years in Japan to produce sake, miso, and soy sauce. *Aspergillus oryzae* has been domesticated from an ancestral species that is considered an agricultural pest, *Aspergillus flavus. Aspergillus oryzae* forms white filamentous colonies that cover

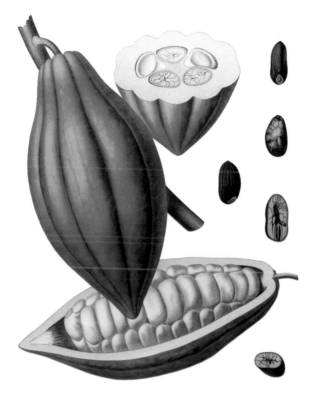

everything, including rice, barley, and cured meats. It has lost, among other un-desirable traits, its ability to produce the carcinogenic toxin aflatoxin.

Chocolate is produced from the seeds of the *Theobroma cacao* plant.

If these examples are not sufficient to prove just how much we should appreciate our food-manufacturing microbes, there is always dessert. Back in the kitchen, the student daydreamer awoke from her reverie and opened a package of chocolate-covered espresso beans. It may not seem obvious that microbes are necessary for the production of two of our most treasured foods, coffee and chocolate. The production of cacao beans and coffee beans begins with a natural fermentation involving many different types of microbes over several days. In the case of cacao beans, fermentation creates the very flavor of the chocolate that we love.

Chocolate in all of its forms begins with cacao pods of the *Theobroma cacao* tree. These pods are the fruit of the tree; they contain seeds we know as cacao beans, even though they are not true beans. When the pods are growing, they

(Opposite) Conidia spores release from *Aspergillus oryzae* conidiophores.

are green, and when they are ripe for picking, they turn yellow. The pod is then split open with a hammer or machete to release the seeds. There are thirty to forty seeds per pod. At this stage they are inedible and covered by a sticky white pulp. It is the fermentation process that makes cacao seeds edible; a trick that human cultures have been using for thousands of years.

Like the chemical residues left behind from fermented alcohol beverages in China, chemical signatures of cacao were recently unearthed and analyzed. Evidence for the early existence of chocolate is spread across Mesoamerica: in 2,600-year-old ceramic jars in Belize, in 3,200-year-old jars in Honduras, and most recently, 3,700-year-old vessels in Veracruz, Mexico. Mesoamerican peoples consumed chocolate in solid and liquid forms. Cacao pods were one of the most valuable items in Mesoamerican cultures, akin to any other form of currency that we have today. Mayans believed that gods shed their mythical blood on cacao during the production process. Aztec peoples thought that one of their gods, Quetzalcoatl, was cast out by the other gods because he shared chocolate with humans. They mixed chocolate with chili peppers and water and poured the mixture between two different vessels over and over, creating a spicy drink with a delicious froth on top. This process was recorded by the Aztecs in the Codex Tudela.

Aztec peoples made a frothy chocolate beverage by pouring the concoction from one vessel to another repeatedly.

For all these years, the critical first step of cacao fermentation proceeded on its own. Seeds were simply piled and wrapped in banana leaves. The microbes came along from the hands of humans, from a waft of air, on the banana leaves, or on the wall of the box, left over from the last batch. The traditional chocolate maker would simply return in a week and, as if by magic, the seeds would be transformed into a palatable product that could be further processed to become chocolate.

What really happens within the pile of fermenting cacao seeds? First, the pectin-rich white seed pulp is colonized and broken down by yeasts. The yeasts are first on the scene, breaking down sugars in the oxygen-poor center of the pile and growing in number to dominate the seeds for the first few days. As they do so, the yeasts produce ethanol, carbon dioxide, and heat as by-products. They increase the pH of the pulp from 3 to a less-acidic 5 as they consume citric acid. And they open up the pile to more oxygen as they create holes in the pectin-rich

(Opposite) The white pulp of cacao seeds is colonized by microbes during fermentation.

pulp. It is this very initial success that causes a crash in the yeast population after about two days. The environment the yeasts created—10 percent ethanol, 45°C, more aerated and higher in pH—kills off the yeast population and also kills the plant embryo within the seed.

As the yeast cell population declines, lactic acid bacteria are able to live in the new conditions on the pulp. The lactic acid bacteria further break down sugars and produce lactic acid and other products. Next, acetic acid bacteria join the process from days two to four. The acetic acid bacteria convert the ethanol that has been produced previously into acetic acid. In the last few days, fully aerobic spore-forming bacteria like *Bacillus* species and filamentous fungi rise in numbers. Just like the microbes on the cheese rind, the microbes that live on the pulp produce many compounds that add a rich and unique set of flavors. All of this happens in more or less the same sequence every time cacao seeds are fermented. It's a process known as ecological succession. Each group of microbes grows on the products of the group that grew on the pulp before it. Ultimately, between five and seven days after the process begins, fermentation must be put to a stop before the microbes begin to break down the seed itself. At this point the beans are typically dried, roasted, and preserved as ground cacao nibs.

Nowadays, many chocolatiers precisely control the cocoa production process at every step, from bean to bar. Like the grapes used to make wine, raw cocoa seeds are in many cases inoculated with particular domesticated strains or propagated strains of wild yeasts with the most desirable qualities for fermentation. Using domesticated strains helps to avoid off-flavors that can develop during a spontaneous or traditional fermentation process, with the trade-off of limiting other potentially pleasant emergent flavors.

Over the last few centuries, we have taken the domestication of microbial species to a greater level of precision. Using state-of-the-art genetic tools, we can precisely engineer strains of yeast, strains of *Aspergillus,* lactic acid bacteria, and other microbes to achieve our preferred ends. Certain microbes have become even more adaptable and versatile than centuries of old-fashioned artificial selection made them. *Aspergillus* cells have been genetically engineered to produce the proteolytic enzyme chymosin, previously needed for cheese curd production, so it is no longer necessary to use rennin obtained from a calf's stomach.

Our technologies are the utmost examples of horizontal gene transfer on Earth. The scientific field of synthetic biology has given us new powers as the ultimate bioengineers—powers to mix and match traits and delete genes or move virtually any gene from any organism into any other organism, no matter how distantly related the two species are. Yet in doing so, only the engineering is new; only the accuracy and intentionality in our control of biological systems is different. By developing new ways to manipulate genes, we are simply emulating processes that have underpinned nature for billions of years. The entire tree of life on Earth is not a tree: it is a web.

We have seen how, as species pass down genetic information vertically, from generation to generation, they can also exchange genes with each other horizontally, from cell to cell of the same generation. This horizontal gene transfer happens so frequently that it has a real and lasting impact on the shape of evolutionary groups of organisms over time. It occurs between cells of the same species and between cells from distantly related species. We can add our scientific tools to the thick network of horizontal connections between organisms that create genetic diversity and innovation within the great web of life.

One person's daydream about a rind of cheese has led us to insights into the evolutionary process itself. This is not evolution in deep time, the kind of evolution tracked over millions of years and recorded in rocks. This is evolution in real time. It is evolution that happens through the rapid succession of generations and through horizontal gene transfer. We can watch new microbial strains arise before our eyes; we can see how those strains come about by looking at the exact changes within the genome; we can understand how those changes manifest and help the organism adapt to its environment. In microbes, we can track evolution at the molecular level: the fundamental process of life.

7

There Is Life at the Edge of Sight

ON AN OCTOBER EVENING IN 1923, thirty-three-year-old American astronomer Edwin Hubble looked up at the Californian sky. The arm of our Milky Way galaxy began to appear as the sun set over the San Gabriel Mountains northeast of Los Angeles. He was eager to get to work. Hubble had access to the shiny new Hooker Telescope, the largest telescope in the world from 1917 until 1949. The telescope's two-and-a-half-meter mirror collected more light and could see farther into space than any other instrument in the world. He was on the verge of a breakthrough.

In the stillness of that night, Hubble tilted the telescope toward the constellation Andromeda. There was the Great Andromeda Nebula: a celestial object that looked like a blurry star to the naked eye but was spiral shaped when viewed with a telescope. He brought the nebula into perfect focus and began a forty-five-minute exposure on a photographic glass plate. As the chemical coating on the glass reacted with light reflected by the telescope, little black dots developed within the haze. That plate of glass, labeled H335H, was the 335th image that Hubble captured using the Hooker Telescope. Little did he know that it contained one of the most important images in astronomy.

Hubble grew up in an era when we thought our Solar System existed within an island universe. Most astronomers thought that nothing was detectable past the outmost stars of the Milky Way galaxy. By comparing image H335H with all of his previous images of the Andromeda Nebula, Hubble identified what he first thought was a supernova, labeling it "N." Later, he realized it was not a supernova—it was a special kind of star that increases and decreases in brightness. Ecstatic, he crossed out the "N" and relabeled it "VAR!" for variable star. Hubble knew that if he measured the star's intensity over time, he could determine its distance from Earth. He continued capturing images, and before long, he had all of the evidence he needed to show that the Andromeda Nebula was not a nebula at all. The star he measured was in another galaxy. It was an island universe unto itself. Those black dots on the glass expanded our perception of the natural world. The light that made the dots came from a star that is larger than the Sun.

A band of our Milky Way galaxy as Edwin Hubble may have seen it. When we see the Milky Way Galaxy like this, we are looking at the edge of the galaxy from a position within it. Thus we see a band of stars that, in very dark locations, forms a complete semicircle across the sky.

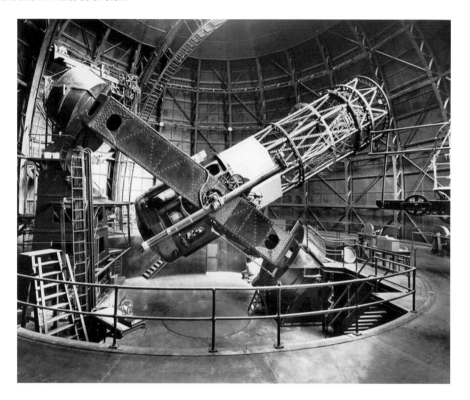

The Hooker Telescope at the Mt. Wilson Observatory in California. Edwin Hubble worked at the observatory in 1919 as a staff scientist shortly after the end of World War I. He had enlisted in the US Army but was never deployed. Hubble was uniquely prepared for his work with this telescope, having completed his PhD dissertation entitled "Photographic Investigations of Faint Nebulae" in 1917. He remained at Mt. Wilson from 1919 until his death in 1953.

The star is so far away that the light it emitted, traveling at 299,792 kilometers per second, took 2.5 million years to reach Earth and Hubble's glass plates.

Edwin Hubble's spirit for exploration lives on through one of mankind's most successful scientific endeavors: the Hubble Space Telescope. Launching a telescope into orbit around Earth eliminates the interference created by the atmosphere, one of the biggest limitations of ground-based telescopes. Since 1990, the Hubble Telescope has been capturing stunning images of the cosmos and sending them back to Earth.

The Hubble Extreme Deep Field is one of NASA's best snapshots of the universe. A Deep Field image shows over 5,000 galaxies hidden in a tiny patch of sky that to the naked eye appears dark and featureless. There are elliptical and irregularly shaped galaxies and spiral galaxies like the Milky Way and

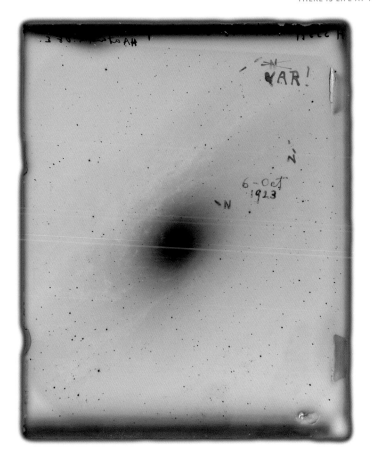

Hubble's glass photographic plate, No. H335H, captured on October 6, 1923. This forty-five-minute exposure photograph was a pivotal point in Hubble's research, ultimately leading him to show conclusively that the object he was looking at was another galaxy.

Andromeda. All of these galaxies are in constant flux, evolving and moving away from or toward one another. When imaged, galaxies moving toward Earth appear blue, and those moving away from Earth appear red. The smallest of the red objects are more than 13 billion light years away. We can't look much farther—light appeared only a few hundred thousand years before this, and the universe itself is 13.8 billion years old. Edwin Hubble broke the limits of the known universe with the Hooker Telescope in 1923, and now the telescope named in his honor has again brought us to the edge of the observable universe.

(Overleaf) The Andromeda galaxy, formerly called the Andromeda Nebula.

Inside of a typical galaxy packed with planetary systems. These are extrasolar planetary systems, depicted around other stars within the Milky Way.

(Previous) The Extreme Deep Field from the Hubble Space Telescope. This image represents light from over twenty-three days of total exposure time, collected by the Hubble Space Telescope. The faintest objects within the image are ten billion times fainter than can be detected by the human eye in the night sky.

Our observable universe is a wilderness of over one trillion galaxies. Each galaxy contains billions to trillions of stars, and most stars are orbited by planets. The Hubble Deep Field image pulls from each one of its viewers the most visceral questions at the core of what it means to be human. Are there other worlds like Earth? Are there other intelligent entities with telescopes observing our corner of the universe? Where within those pixelated red and blue galaxies could there be life?

Astronomy provides the tools to begin the search for extraterrestrial life. We can detect planets in nearby solar systems by turning telescopes to other stars within our own galaxy. In the Milky Way there are billions of alien worlds called exoplanets that might harbor life. We are beginning to fill in the details about these exoplanets with specialized instruments like the *Kepler* spacecraft. The aim is to determine features of exoplanets that are consistent with life, like the abundance of water, the presence of an atmosphere, and a position within the habitable zone from a sun: not too hot and not too cold for life. The Sloan Digital Sky Survey uses a two-and-a-half-meter telescope located at Apache Point Observatory, New Mexico. It is capable of detecting signatures of individual elements near distant stars by collecting infrared light and separating it through a prism. The telescope has detected all of the key elements of life on Earth (carbon, hydrogen, nitrogen, oxygen, phosphorus, and sulfur) in solar systems throughout the Milky Way. A similar distribution of life's elemental precursors is likely found in the solar systems of other galaxies as well because all galaxies exist within the same universe and follow the same fundamental laws of physics. The vast distances between Earth and these exoplanets, of course, even those here in the Milky Way galaxy, precludes getting probes anywhere near them right now.

We cannot fully contemplate whether there is life on other planets and what that life might be like with evidence from astronomy alone. By combining our knowledge of astronomy and biology, however, we can begin to describe some of the essential features of life outside Earth by defining the very basic properties of life here on Earth. This combined field of inquiry known as astrobiology gives us the tools to detect one or more signatures of life in the universe, to find the full range of such signatures, and to ask the weighty question: What is life?

To understand life on Earth, we need a high-resolution view of biological worlds. Lucky for us, as the observable universe expanded over the past few hundred years through astronomy, the observable microverse of biology expanded

(Overleaf) Five Earth-like exoplanets: alien worlds shown next to Earth (*far right*). From the left, the planets are Kepler-22b, Kepler-69c, Kepler-452b, Kepler-62f, Kepler-186f, and Earth. Of these, Kepler-452b stands out as one of the most Earth-like planets discovered so far. Kepler-452b orbits a G2-type star similar to the Sun and has an orbital period similar to Earth's with years lasting 385 days. It is 1.6 times the size of Earth and around 1,400 light-years away.

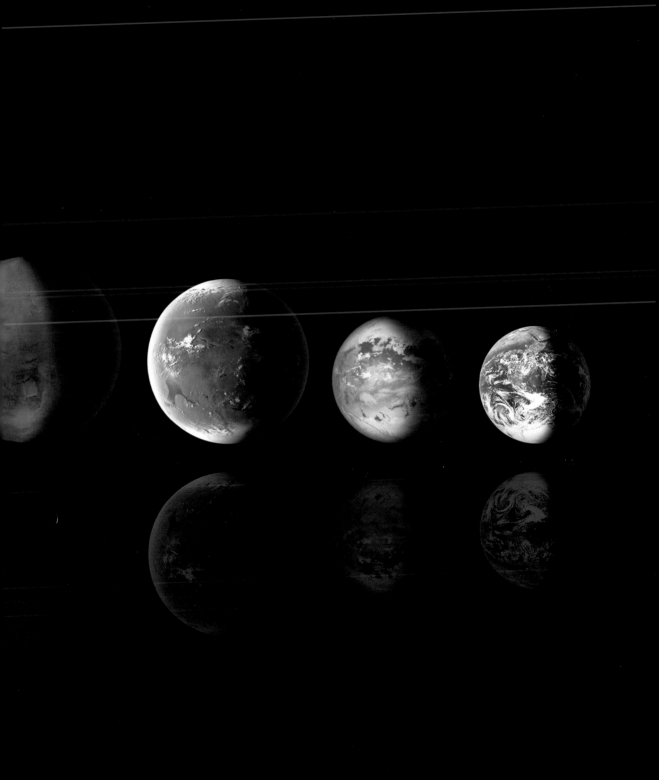

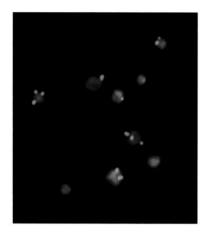

Archaeal cells from Yellowstone seen in red with ultra-small green *Nanopusillus* ectosymbionts. Both organisms are from the domain Archaea. The larger red cells are *Acidilobus*, and the smaller green ectosymbionts attached to the red cells are *Nanopusillus acidilobi*.

as well. The tools in both fields have moved well beyond visible light, capturing data from wavelengths outside of the visible spectrum at size scales invisible to our unaided eyes.

As a consequence, biologists are questioning some of the bedrock concepts of life. Is the human body, together with all of the microbes that live within it, an organism, a metaorganism, a holobiont, or an ecosystem? All of these terms may be used to describe humans. The same question applies to other animals, such as leaf-cutter ants and corals that co-evolved with their microbial symbionts and respond developmentally to them. Zooming to the nanoscale, an ultra-small archaeal cell named *Nanopusillus* lives in the hot springs of Yellowstone National Park in permanent association with a larger archaeal cell. The association is so tight that the ultra-small cell attached to the outside of the larger cell as an ectosymbiont cannot even make its own ATP, one of the basic requirements for life. Is it a separate organism? Or can we think of the ultra-small cell and the host cell as one functional metaorganism? We are comfortable calling biological entities such as single-celled eukaryotes organisms because their ancient microbial endosymbionts have become so naturalized that we've given them new names—the mitochondrion or the chloroplast. Where do we draw the line between mutually dependent symbiotic associations of two or more microbial species and a single organism? These questions in biology arose only after we realized that microbes make up the overwhelming majority of the biodiversity on Earth and after we exposed the connections between larger multicellular eukaryotic organisms and microbes, and among microbes themselves.

The ubiquity of microbes and their foundational role in the evolution of life on Earth leads to the hypothesis that wherever in the universe life originates and begins to evolve, that life will begin with microbes. Life is an emergent property of chemical interactions: it will emerge from chemistry that occurs at size scales too small for us to see. Each independent origin of life will be microscopic simply by virtue of the size of the molecular precursors that interact to form living systems like protocells, even if the precursors themselves are not the same.

This "microbes first" theory of the origin of life on all planets establishes a directionality at the coarse scales of biological evolution: it predicts that on all planets with life, that life will be maintained in a mostly microbial state. From the moment that the first microbes arose on Earth around four billion years ago, they grew and evolved and diversified to occupy every habitable niche on the planet. In doing so, they shaped the planet for the next phase of life. Most animals began evolving a mere 500 million years ago and only because microbes produced the oxygen in the atmosphere first. Terrestrial plants evolved only within a soil created by microbial decomposition. Over the billions of years that it took for more complex life to evolve, microbes invented most of the biochemistry on Earth. They control global biogeochemical cycles and influence climate. Animals evolve from a microbial world, within a microbial world. They are sustained by microbial species, and they require a functional microbiosphere to survive.

Extinction events and other conditions that challenge life also keep planets in a mostly microbial state. The physical forces that cause extinction events are not unique to Earth. All planets are susceptible to such dangers as radiation, comets, and asteroids. If the planets have an active molten core, the consequent thermal activity and volcanic eruptions represent another challenge to life-forms. As planets cool they may transition from nonliving to living and back to nonliving if, for example, the molten core cools and the protective magnetosphere is deactivated. Each potential extinction event in the universe, like the five predicted mass extinctions that have occurred on Earth, has the capacity to readjust ecological and evolutionary trajectories at a planetary scale. Yet during each one of these events, such as the one that killed nonavian dinosaurs, or during hypothesized "Snowball Earth" periods when the planet was covered by glacial ice, microbes survived. Bacteria like *Deinococcus radiodurans* can survive a dose of over 10,000 gray units of ionizing radiation by virtue of sophisticated DNA repair systems. By comparison, just 10 gray units of whole body ionizing radiation exposure causes certain death in humans. There is also a "deep biosphere" on Earth: a reservoir of microbial life kilometers below the surface of the planet. And microbes have mastered survival with a variety of spores and other forms of long-term dormancy. Halophilic archaea, for instance, can survive encapsulated within fluid inclusions inside salt crystals for millennia.

While thinking of microbes on Earth in such a wide context is thus particularly useful for speculating about exotic biologies that could exist on the other side of the universe, or in our cosmic neighborhood, it also leads us to question

Halophilic archaea can survive for millennia trapped inside halite salt crystals. The species shown is *Haloferax volcanii*, visible within the salt crystal because of its natural orange carotenoid pigments.

the very concept of a microbe. A microbe is an operational definition for the sensory organ of one species on Earth. It's human-centric. Think of it this way: if an organism on another planet has eyelike organs that capture light like ours but can resolve objects much smaller than we can, then our microbe may be their macrobe. The same may be true of organisms on Earth with sharper visual acuity than humans. Is the definition of a microbe entirely relative?

The truth is, there is no real division between the microbial world and the rest of nature. Every organism in the biosphere exists within a continuum that includes all living systems. Consider the size range of a variety of different organisms. There are giant bacteria that dwarf the smallest known animals. One giant bacterium, *Thiomargarita namibiensis,* hails from ocean sediments off the coast of Namibia and forms individual cells about the size of the white space within this letter "o." Single cells of the bacterium *Epulopiscium fishelsoni,* which live as a symbiont within the intestines of surgeonfish, can also get that big. These bacterial cells are larger than the smallest crustacean, the smallest insect (the 134-micrometer parasitic wasp, *Dicopomorpha echmepterygis*), and the smallest flowering plant (a duckweed). Multicellular filaments and spheres of some *Thiomargarita* bacteria are even larger than the smallest vertebrates, including the dwarf pygmy goby and a tiny frog from Papua New Guinea.

In addition to giant bacterial species, many microbial species form macroscopic structures during their natural life cycles or lifestyles. These include the fruiting bodies of fungi (mushrooms), fruiting bodies in myxobacteria, aerial mycelia in *Streptomyces,* and the plasmodial feeding stages and fruiting bodies of slime molds. Natural macroscopic manifestations may develop in certain environmental conditions. There are, for example, the macroscopic biofilms formed by most bacteria, archaea, and many microbial eukaryotes. The greenish fuzz seen on any common river rock is the macroscopic display of millions of tiny glass diatoms. This category also includes microbial mats that contain many species, like those seen in Yellowstone National Park and in salt marshes. Microbial networks are often macroscopic too, like the "humongous" honey fungus in Oregon, which is larger than some human cities.

Finally, there are passive or loosely associated microbial agglomerations. Microscopic bacteria, archaea, algae, and other microbial cells have the capacity to grow at high density in a liquid in the right conditions, becoming visible in natural settings through turbidity, particularly when they are naturally

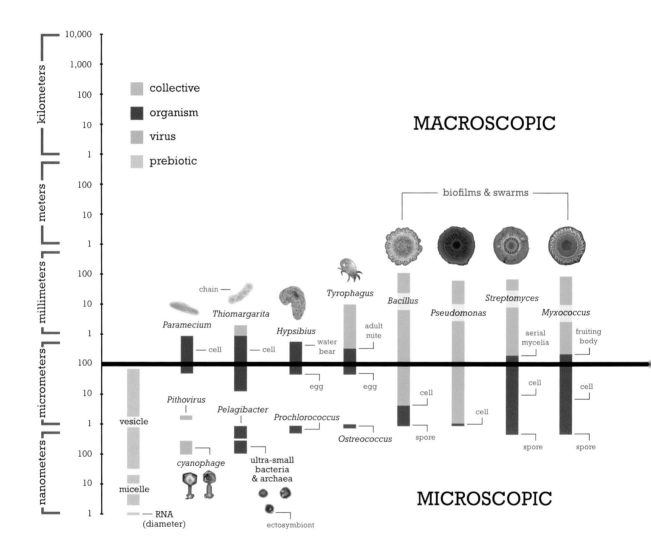

10,000
1,000
100
10
1
100
10
1
100
10
1
100
10
1
100
10
1

kilometers
meters
millimeters
micrometers
nanometers

collective
organism
virus
prebiotic

MACROSCOPIC

biofilms & swarms

Thiomargarita

chain

Paramecium

cell

cell

Hypsibius

Tyrophagus

water
bear

adult
mite

Bacillus

Pseudomonas

Streptomyces

Myxococcus

aerial
mycelia

fruiting
body

cell

cell

spore

spore

egg

egg

cell

cell

spore

vesicle

Pithovirus

Pelagibacter

Prochlorococcus

Ostreococcus

micelle

cyanophage

ultra-small
bacteria
& archaea

RNA
(diameter)

ectosymbiont

MICROSCOPIC

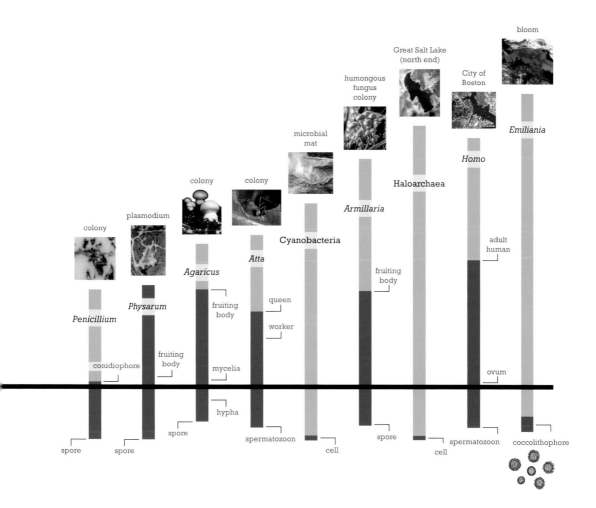

The size scales of life. The size range for twenty-six biological entities is shown on a logarithmic scale. The entities are divided into categories of prebiotic chemical entities (like membrane vesicles), viruses, cellular organisms, and collective forms of organisms. For each organism, dark green bars show the strict life cycle or natural lifestyle—the range in sizes for the biological forms involved in that organism's reproduction. For example, the life cycle of *Agaricus bisporus* spans between microscopic spores and mushroom caps that complete the life cycle by releasing more spores. Light green bars are collective forms of those organisms; for example, a biofilm or microbial mat is a collective form of bacteria, a bloom is a collective form of algae, and an ant colony is a collective form of leaf-cutter ants. The dark, horizontal line is set to the visual acuity of the human eye, able to see objects about 0.1 millimeters and larger (when very near to the observer). This establishes the general rule that everything below the line is microscopic and everything above is macroscopic.

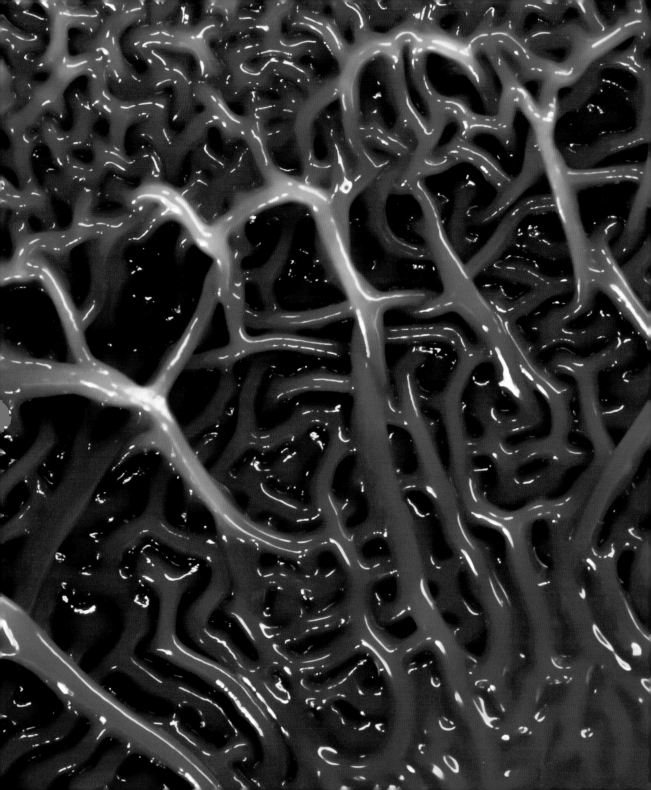

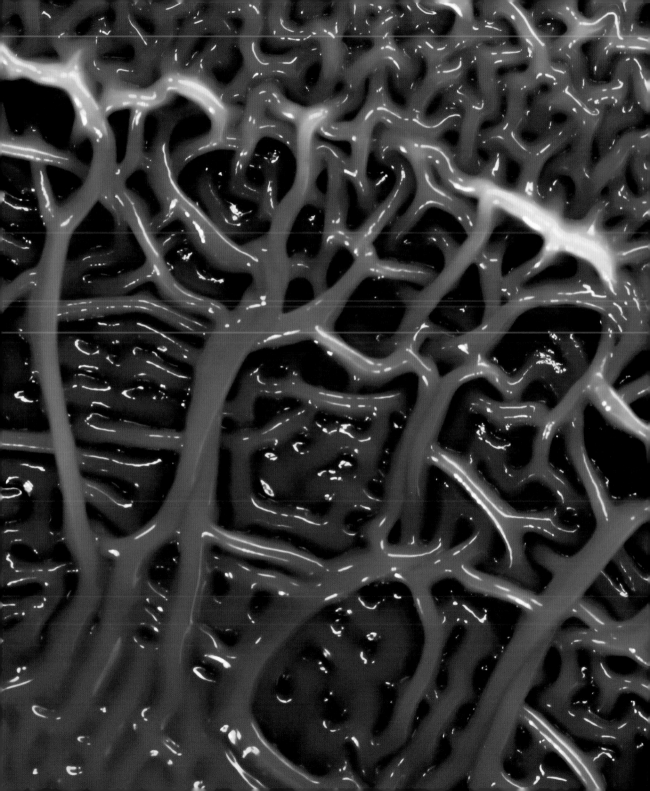

pigmented. This is how haloarchaea with their carotenoid pigments turn the entire north half of the Great Salt Lake deep red. It is also how cyanobacteria turn ponds and seas bright green, and how thermophilic microbes color the beautiful landscapes of Yellowstone. And we cannot forget the little coccolithophore alga *Emiliania huxleyi,* which blooms to become one of the largest biological forms on the planet.

So far we have considered the macroscopic manifestations of organisms usually considered microbes, but we can also look at this issue by considering the microscopic stages of creatures that we never think of as microbial. In reality, the biology of every large multicellular plant and animal is rooted in the microscopic scale. Animals like humans are visible because they spend the majority of their life cycle as cells organized into macroscopic tissues, organs, and bodies. Nonetheless, we are entangled with the microbial world. At one moment, the fate of Edwin Hubble and of every other human that has ever lived was carried forward by a single sperm cell five micrometers wide that fertilized an egg cell whose size is right at the edge of sight. The same is true of great blue whales. Likewise, the tallest trees depend on microscopic pollen to complete their life cycle.

In light of the complexities of the life cycles and lifestyles of organisms from all domains of life, we should really designate a new term: *eumicrobe.* A eumicrobe is any organism or biological entity whose entire life cycle or lifestyle is microscopic, lacking in any natural macroscopic manifestations. By this definition, other microbial life-forms that are sometimes microscopic and sometimes macroscopic are "facultative microbes." When we apply this view of life, we find that many organisms taken for granted as microbes are not eumicrobes. Bacteria, a group treated as synonymous with the word *microorganism,* are not categorically microscopic because of giant single cells and because of the many macroscopic manifestations of other bacterial species. The *Archaea* do not qualify. There are archaeal filaments that grow up to three centimeters long on sunken wood in mangroves of the French West Indies, made of cells up to twenty-five micrometers wide. There are also natural archaeal biofilms that can be seen with the naked eye, like the biofilms of haloarchaea seen on the shores of Great Salt Lake and a "string-of-pearls" type biofilm found in a cold sulfurous marsh in Germany.

(Previous) Biofilms are macroscopic manifestations of microbes. This biofilm was formed by the bacterial species *Pseudomonas aeruginosa* on solid growth medium.

(Opposite) The colorful scenes of Yellowstone National Park are macroscopic manifestations of thermophilic microbes.

So what does this leave us with? Who are the eumicrobes? Even though macroscopic microbial forms are common, many bacteria and archaea certainly are eumicrobes. There are many parasites of animals like *Trypanosoma* and *Plasmodium* that are always microscopic. The marine bacterium *Prochlorococcus* also qualifies. *Prochlorococcus* cells reproduce through a basic life cycle of binary fission and range from 600 nanometers to 800 nanometers wide; they grow to visibly high density in the laboratory, but do not grow as large blooms in nature. The other tiny bacterial species that *Prochlorococcus* interacts with in the ocean, *Pelagibacter,* qualifies as a eumicrobe as well. Other examples are bacterial cells from the genus *Mycoplasma* that are 300 micrometers wide and have some of the world's smallest genomes for independently reproducing cells, encoding less than 500 genes. Ultra-small archaea and bacteria species can have even smaller genomes and cell sizes than *Mycoplasma*. This is because, as symbionts, they have lost many functions over time that became redundant in the context of symbiosis. These are species like the archaeal ectosymbiont *Nanopusillus* encountered in the Yellowstone hot springs, and the candidate phyla radiation of bacteria, ranging from 100 nanometers to 300 nanometers. There are even some animals that are eumicrobes, like *Tantulacus dieteri*, a parasitic crustacean with an adult size of just 85 micrometers.

Viruses are the only group of biological entities that are categorically microscopic. Viruses range in size from well into the nanoscopic realm to the giant viruses like *Pandoravirus* and *Mimivirus* that are larger than many bacteria. One virus resurrected from a 30,000-year-old permafrost in Russia, named *Pithovirus,* is 1.5 micrometers wide. That makes it larger than the smallest free-living eukaryote, the green alga *Ostreococcus*. And yet, as the only categorically microscopic group of biological entities, they are just that: biological entities. Viruses lurk outside of the web of cellular life, moving in and out of the genomes of cellular organisms. But they are not full-fledged organisms themselves because all viruses we've found so far are dependent on other species to reproduce and lack some of the other perquisites needed to be considered fully "alive." One might even consider small viruses like bacteriophage viruses as little more than glorified biomolecules. However, viruses clearly have emergent properties that the molecules that compose them do not. (Bacteriophages can even communicate with each other through small molecules in the environment, reminiscent of how bacteria communicate through quorum sensing.) So, rather than debate whether or

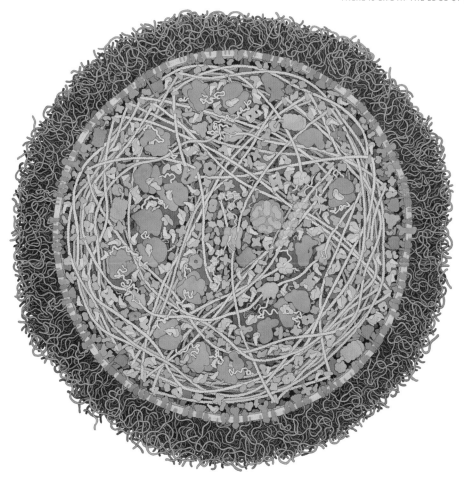

not viruses are alive, we can agree that viruses are eumicrobes, and that they simply exist as gray tones within the spectrum from chemistry to biology that defines life on Earth.

Extraterrestrial life could exist at any of the size scales of biological organization that we know exist on Earth. Imagine now that we take all that we know about life on Earth and fix the Hubble Space Telescope on Europa, one of the moons of Jupiter named by Galileo in the early 1600s. Europa is about 10 percent smaller than Earth's moon, and it is covered by a solid layer of ice.

An artistic view inside a *Mycoplasma* cell, 300 nanometers wide. The species shown is *Mycoplasma mycoides*, a parasite of cattle, goats, and other ruminants. The model is composed of many structural and functional molecules.

Suspected water plumes captured by the Hubble Telescope. The bluish-black background is an ultraviolet image. The image of Europa superimposed on the ultraviolet image was collected by the *Galileo* and *Voyager* spacecraft. The features predicted to be water plumes (at the seven o'clock position of the Moon) are estimated to be 160 kilometers in height.

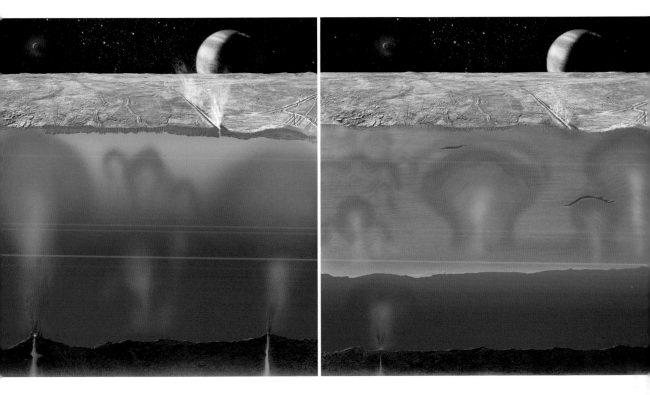

Water plumes originate from an ocean created by thermal activity beneath the icy surface of Europa. Two possibilities for the subsurface of this moon are shown. One possibility is that thermal activity within the rocky layer below is intense and directly melts the ice nearly all the way to the surface. It is also possible that thermal activity is not as intense, creating a relatively thin layer of liquid water, which then transmits heat and some liquid water to the surface through a thick layer of ice, explaining the features and activities seen on the surface. Jupiter, thought to be the ultimate cause of most of this thermal activity due to gravitational interactions, is seen in the distance.

The presence of all of this water is one of the reasons that the astronomer Carl Sagan short-listed Europa in 1971 as a likely site of life in our Solar System. The telescope spots white blasts coming from the icy surface of Europa that look like colossal water plumes. If that is liquid water, where did it come from? Could the water be alive with microbes? As Europa orbits Jupiter once every eighty-five hours, gravitational interactions with Jupiter cause tidal flexing forces comparable to plate tectonics on Earth. This tidal flexing creates heat and geothermal activity underneath the ice shell and explains Europa's heavily

(Overleaf) The ice-covered, cracked surface of one of Jupiter's moons: Europa, based on data collected during fly-bys of Europa by the *Galileo* spacecraft in orbit around Jupiter. The white and blue areas are ice, and the red areas are other geological features containing less ice.

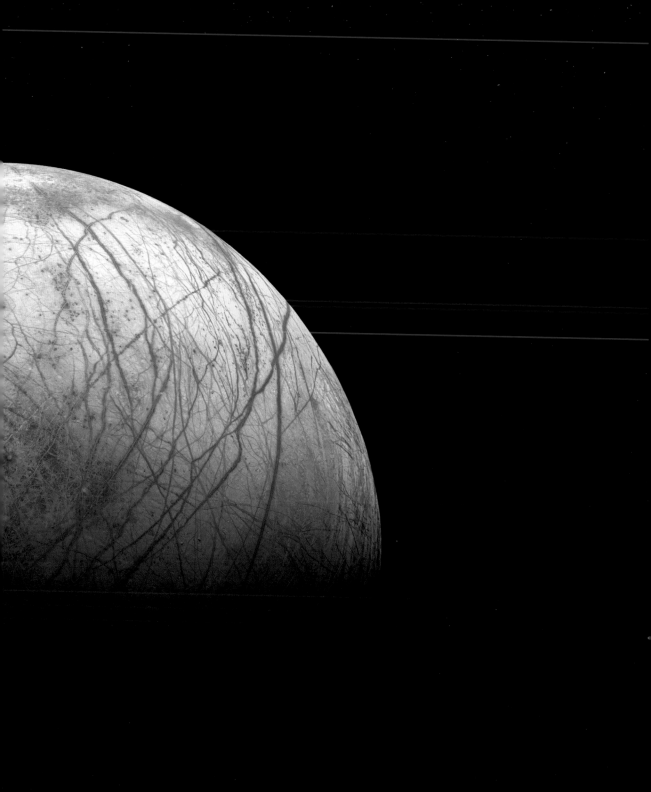

The next-generation James Webb Space Telescope. The 6.5-meter primary mirror of the telescope is composed of eighteen hexagonal segments that will unfold once launched into space. The mirrors are made of beryllium, coated with gold. In addition to having a larger mirror than the Hubble Telescope, the James Webb Telescope will orbit farther away from Earth, and it has a large sunshield, allowing it to maintain a stable environment ideal for imaging.

cracked surface. An artist's impression of Europa shows two different models of how the resulting geothermal activity could create either a thin or a very deep unseen ocean below the ice. Either way, measurements from the *Galileo* spacecraft indicate that there is liquid water and that this water is likely very salty. All of the conditions seem right for life. Soon the James Webb Space Telescope will launch as the successor of the Hubble Telescope. What will this telescope, with its 100 times more powerful 6.5-meter gold-coated honeycomb mirror, see on Europa? That is a physical aperture for capturing light exactly 1,000 times wider than the maximum aperture of the pupil in the human eye. On Europa, on Saturn's moons Titan or Enceladus, or perhaps on a more distant exoplanet, there is a good chance we will find clear evidence of extraterrestrial life in the next few decades. Several space missions to Europa are planned for the 2020s, including NASA's Europa Multiple-Flyby mission.

Hubble's images opened the cosmos and in so doing opened the possibility of extragalactic life. The more cosmic real estate there is out there, the greater the mathematical likelihood for an origin of life event and for advanced civilizations to develop. Our island universe, once only the Milky Way galaxy containing some 100 billion planets, is now a universe filled with a total of some septillion planets. That's 10^{24} planets, ten trillion more places where life might exist.

We are making beautiful maps of the largest and the smallest structures in the universe. We are learning how galaxies themselves interact to form galaxy clusters and how galaxy clusters interact to develop into a lattice of threadlike filaments made of dark matter. It's called the cosmic web. Galaxies form along filaments and cluster at the nodes in the web where filaments join. Just as we could never see a virus by looking at individual atoms alone, just as we could not see a biofilm by looking at individual bacterial cells, we could never see the cosmic web by looking at individual stars and galaxies. Simultaneously, we are exposing how analogous networks of neurons develop in the brain. And we have found at the apex of many scientific disciplines that on Earth, everything depends on everything else. How could we ever take a picture of that complete

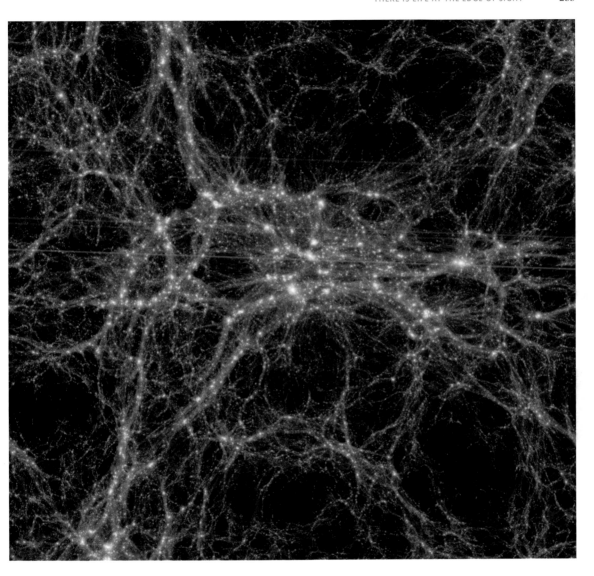

The cosmic web: a large-scale structure of the universe that evolves over time. This is an image from the Millennium-II computer simulation, a study that set out to model the distribution of matter across a hypothetical cube of the universe as a way to understand how large-scale structures like galaxy clusters and galaxy filaments came about. The simulation was populated by millions of galaxies; it required 1.4 million hours of computation divided over 2,048 processing cores to produce. The image represents a section of the simulated universe 326 million light years wide.

interdependency? What would the image look like, if we captured in one large-scale view, like the cosmic web is for the universe, the sum of all of biological activity across the web of life, all of the connections between organisms in ecosystems, humanity, the forests, the living ocean of microbes, the living soil, and of the ways that those ecosystems impact global patterns? We already have this image. It's Earth.

There is life at the edge of sight in the microbial world on Earth that together creates structures and processes that are anything but invisible and anything but small in impact. There is life at the edge of sight in those points of light in the night sky traveling from stars that hide trillions of alien worlds and almost certainly alien life. Life at the edge of sight is the recognition that nature is seamless and that all organisms blend within interconnected living systems irrespective of our best scientific definitions and despite how hard we try to divide and categorize them. Microbes gave us life. Just as we see beauty in the rainforests and oceans and in the cosmos, so should we see, celebrate, and continue to explore beauty in the microbial world.

Composite image of Earth's western hemisphere based on images of the land surface layer, the sea ice layer, the ocean layer, the cloud layer, city lights, and the topography layer.

HOW TO PHOTOGRAPH MICROBES

There are two sides to every scientific image. There is the instrument or method used to produce the image, such as the magnifying lens of a microscope. And there is the recording and preservation of that image onto an analog or digital medium. The first microscopes capable of producing images of bacteria and other individual microbes (like Antoni van Leeuwenhoek's) were based on a single glass element that bent light to magnify objects. The first recording media to match these simple light microscopes were the observers themselves, together with pen and paper. From those early days in the seventeenth century onward, light microscopes (also called optical microscopes) improved to achieve sharper and more consistent images at higher magnifications, while the common recording medium remained pen and paper for centuries. Far from being a period of crude drawings, it was a remarkable time when scientific discoveries and scientific education was driven by sometimes masterful illustration.

Then, a major change came in the nineteenth century with the invention of photography. Early versions of photographic cameras were immediately adapted to microscopes (and to telescopes). Microscopy and photography were now advancing in sync: sharper, more magnified images from microscopes were captured by cameras that preserved these images with greater clarity and convenience. Today, we have the latest in light-based microscopes and macro lenses, mated to high-resolution digital image sensors sensitive to the smallest

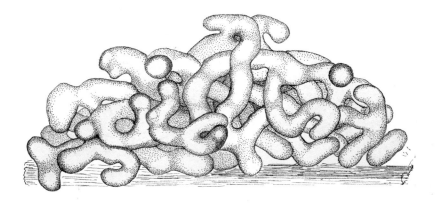

(Opposite) Green algae from the genus *Pediastrum*, drawn by Ernst Haeckel in 1904.

A macroscopic structure formed by myxobacteria from the genus *Archangium*, drawn by Roland Thaxter in 1892.

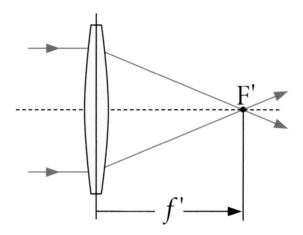

Light refraction through a simple glass lens. This is a basic diagram showing a beam of parallel light rays (a state known as collimated light) traveling from the subject toward a glass lens. The light rays are bent (also known as refracted) by the convex piece of glass, causing the rays to converge upon a focal point (F′) at a fixed distance from the lens, known as the focal length (f′). The focal length is determined by the shape and other properties of the lens.

amount of light. We have fluorescence microscopy and confocal microscopy that allow for labeling of many structures or cell types simultaneously and in three dimensions. And twentieth-century technologies such as electron microscopy have brought our scientific viewpoint beyond the limitations of visible light and well into the nanoscale. The following sections describe the different types of images, beginning with landscape and macro photography to light microscopy, fluorescence microscopy, and confocal microscopy. Our final subjects are scanning electron microscopy and even more advanced methods not seen in this book.

Photographing the Macroscopic Manifestation of Microbes

Many of the images in this book of microbial science are standard 35 mm photographs captured with a digital single-lens reflex camera, or DSLR. We have emphasized the macroscopic manifestations of microbes, elaborated in every chapter but particularly in chapters 4 and 7. The value *35 mm* is the size of the image sensor (36×24 mm; also called full-frame) that converts light information within an image into digital information inside of the camera: light that was collected and focused onto the sensor through a system of glass elements within the camera's front-facing lens. While one simple convex glass lens can produce a somewhat focused image, the many optical elements within modern camera lenses work together to correct optical aberrations like image distortion,

astigmatism, and spherical effects. Optical aberrations are similar to the wonky distortions seen when looking at text through a handheld magnifying lens (an example of a simple lens). As an imaging technology, modern DSLR cameras trace their pedigree to the first DSLRs in the 1990s and to the first 35 mm film SLR released by Leica in 1935. Film cameras trace their pedigree all the way to the daguerreotype process of capturing images on metal sheets lined with light-sensitive chemicals. These were the first mainstream photographs, introduced in 1839. And daguerreotypes owe their existence to the ancient pinhole image, or camera obscura.

One key advantage of DSLR camera design is the ability to interchange lenses of different focal lengths. Focal length (f') is the length between the outermost lens element and the focal point: the point at which the captured refracted light rays converge to produce a focused image (F′ in the diagram). Focal length is one of the basic properties of an optical system. In practice, switching between lenses with different focal lengths allows the photographer to control

A Yellowstone Mud Volcano, frozen by a very short exposure time of one two-thousandths of a second. The "mud" in this thermal feature approached 90°C.

Long-exposure photograph of a Yellowstone Geyser field at night.

(Opposite, above) An alien microbial landscape of liquid droplets on a fungal colony isolated from the environment. Water droplets like these contain metabolic by-products produced by the microbial cells as the colony grows.

(Opposite, below) Migrating bacteria. These bacteria were isolated from soil.

both the field of view (from wide angle to zoomed in) and the magnification of the image (how much larger a given subject is relative to its real size).

The images of landscapes that set the story in many chapters were captured at wider angles with various camera apertures and exposure times. Lenses used for these wide or intermediate-sized scenes had focal lengths ranging from 14 mm to the more narrow 24, 35, and 50 mm lenses. Exposure time, or the amount of time that the camera sensor was exposed to light (also called shutter speed), varied from image to image from hundredths or thousandths of a second to many seconds. Thousandths of a second of exposure creates still images that appear to freeze the action of a scene, like an exploding pocket of gas within a mud volcano in Yellowstone National Park. Other images, like the night sky in Yellowstone, required up to thirty seconds of exposure. Edwin Hubble used long exposures in the 1920s to capture images of galaxies, and photographers still use long exposure times to collect light from faint or far-off sources. At each of these exposure times, there is a separate aperture setting that controls how much light enters the lens (also called an f-stop). As a general rule, wider apertures (lower f-stop numbers) are good for low-light conditions but

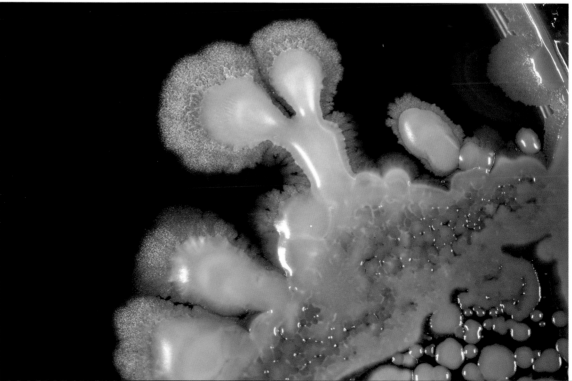

result in less sharpness and less depth of field, which defines how much of the image is in focus—also called focal range. Narrow apertures typically produce images with greater clarity and depth of field, yet images with these higher f-stops require much more light within the scene itself.

Beyond landscape photos, the majority of the DSLR images in the book were captured through macro lenses. Macro lenses have a magnification factor that is at least 1x or greater, capable of projecting a life-sized image of the subject on to the image sensor (also called 1:1). So, for a macro lens at 1x magnification, if the subject is a close-up one centimeter wide of a microbial colony covered in minute liquid droplets produced by the microbes as they grow, there will be, at the time that the image is recorded, a one-centimeter-wide real image of that microbial landscape projected onto the sensor. Macro lenses are tuned for photographing objects that are right around the edge of human sight. These are subjects that are just barely invisible to the naked eye, from cheese mites and other micro-animals to objects about the size of a coin, such as bacterial colonies. Some macro lenses magnify images at up to about five times their actual size (5x magnification), more or less equal to low-power light microscopes such as stereo-microscopes (sometimes called dissecting microscopes). And yet for all of their usefulness, the best macro lenses and stereomicroscopes will never resolve detail at the cellular level for most microbial species.

Photographing the Microscale Realm with Light Microscopy

Photographic technology lagged behind microscopy, so it was many years after Antoni van Leeuwenhoek first drew his animalcules that the world's first photograph was taken through a microscope. Who produced this first photomicrograph is a curious question. Thomas Wedgwood, the uncle of Emma Wedgwood Darwin, wife of Charles Darwin, happened to be one of the first people to propose the idea of capturing a photograph through a microscope. Wedgwood was one of the earliest pioneers of photography itself; he was experimenting in England with light-sensitive chemicals and their application to the camera obscura during the late 1790s when Louis Daguerre, inventor of the daguerreotype, was still a young child living in France. Wedgwood wrote a paper in 1802 together with his colleague Sir Humphry Davy, the chemist who first isolated potassium, sodium, calcium, and several more metals. (Davy was also the inventor of one

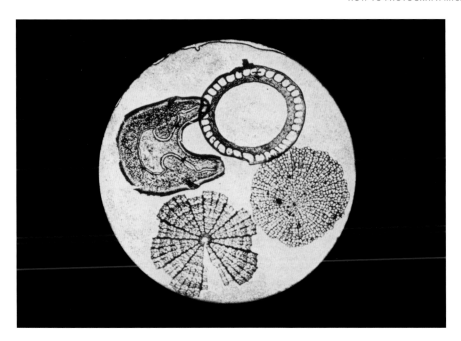

of the first incandescent light bulbs through experimentation with these new-found metals.) In "An Account of a method of copying Paintings upon Glass, and of making Profiles, by the agency of Light upon Nitrate of Silver, Invented by T. Wedgwood, Esq., with Observations by H. Davy," they write: "In following these processes, I have found that the image of small objects, produced by means of the solar microscope, may be copied without difficulty on prepared paper. This will probably be a useful application of the method."

No evidence survives from Wedgwood and Davy to prove that they successfully recorded the envisioned photomicrographs. If they did, they would have used an indirect method, projecting the image from the solar microscope (with the subject illuminated by sunlight) onto a wall or a screen, and then capturing that image using one of Wedgwood's early camera devices. One possible reason for the absence of evidence was the impermanence of Wedgwood's process. The method did not leave the photographs in a fixed state. Instead, unless they were kept in complete darkness, the photographs would continue to interact with light and eventually develop to solid black.

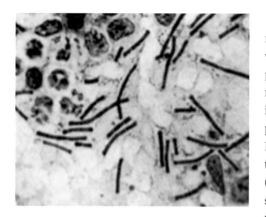

One of the first photomicrographs of bacterial cells by Robert Koch, 1877. The species that Koch photographed here was *Bacillus anthracis,* the causative agent of anthrax.

The first person to successfully apply a photographic method to microscopy was another British scientist and inventor, William Henry Fox Talbot. Talbot invented several paper-based photographic processes capable of producing reasonably permanent photograms or "photogenic drawings" in the 1930s. He combined these processes with lower power solar microscopes that magnified up to about 20x. Beginning in 1839, Talbot captured images of cross sections through plant stems and other similarly sized subjects (such as insect wings). It showed more detail than could be seen with the naked eye, such as the vascular system that transports water and nutrients through the plant, but it was not yet a method that could capture individual cells or microbes.

One of the first people to photograph a microbe in earnest was the French scientist Alfred François Donné. Beginning in 1840, Donné matched the then new and superior photographic process of daguerreotypes (relative to Talbot's prints) with microscopic observation. He combined daguerreotypes with one of his own inventions: the first photoelectric projection microscope, developed in collaboration with one of his students, Léon Foucault. Donné and Foucault's microscope used electric light rather than solar light. With this setup, Donné captured photographs of the ten-micrometer-long pathogen *Trichomonas vaginalis* and of many other microbes, publishing them together in his major work, *Atlas,* a tour of microscopic biological forms including eighty highly magnified images.

In the latter half of the nineteenth century, a number of microscope setups equipped with cameras were developed by German companies and continuously improved upon. Both horizontal and vertical camera microscopes were sold in the late 1800s and into the early 1900s, incorporating such improvements as a connection between the camera and the microscope lens that eliminated vibration. It was a vertical apparatus of this era that Robert Koch used to produce the first photomicrographs of bacterial cells, published in 1877. That same year, one early microscope company, Leitz (which later became Leica Microsystems), sold its 10,000th microscope. Thirty years later, Leitz gave Robert Koch their 100,000th microscope as a gift, just two years after Koch received a Nobel Prize in 1905 for his work on tuberculosis. Although horizontal setups were initially preferred for their stability at higher magnifications, vertical setups were soon made more

The Leitz Panphot microscope of the mid-twentieth century had many of the design features of modern microscopes.

stable and became the gold standard of photomicroscopy and of microscopy in general. In the Leitz Panphot microscope model launched in 1953, we can truly see the beginnings of the modern microscope-camera design used by scientists today.

Modern optical microscopes have a few main illumination modes. Bright-field microscopy is the most basic of these. Bright-field relies on a white light source, like a halogen bulb, that is transmitted evenly through the sample through a method first developed in 1893 called Köhler illumination. Light from the illuminated sample stage is captured by an objective lens that magnifies an image of the subject. The magnifying lenses of today's optical microscopes resemble modern camera lenses in that they contain many optical elements that work together to minimize aberration. The contrast in a bright-field image is created as the light is absorbed more within thicker parts of the sample relative to the bright background. Another illumination mode requiring a slightly more complicated light path is called dark-field microscopy. Dark-field microscopy excludes the light that does not interact with the sample, creating an image composed of only the light that is scattered by the structures under observation. Dark-field images appear inverted relative to bright-field images: the background that would be white in a bright-field image is dark or black, and the structures themselves are bright.

These most basic forms of optical microscopy will always be limited by the translucent nature of most biological samples. One way to get around this limitation is to stain the sample to create more contrast than is naturally present. Staining as applied to microscopy was an important strategy developed in the early days of microbiology. It is still very much at work in its basic and more advanced forms today. There are also modern imaging solutions that create more contrast by manipulating light within the microscope's optical system. Two examples of standard techniques for increasing contrast are phase contrast microscopy and differential interference contrast microscopy or DIC. Specimens under phase contrast microscopy are surrounded by bright halos and objects imaged through DIC have more clearly defined and three-dimensional edges.

The next major step to increase the complexity available in a light-microscopy image is to tease apart a biological subject. If, for example, the subject is a bacterial cell, the individual parts of the cell, from the cell wall to individual proteins, will be indistinguishable one from the other, even if we create more contrast within the entire cell. Classic techniques like the Gram stain, developed in 1884, differentiate based on the interaction between the dye crystal violet with the peptidoglycan in a bacterium's cell wall. Other stains label other biological structures in different colors that can then be seen through bright-field microscopy. There is a better way, however, to differentiate parts of a cell or different cells within a mixed population of cells using another adaptation of optical microscopy: fluorescence microscopy.

While conventional light microscopes use a broad spectrum of visible light (in the wavelength range of 400–700 nm) to produce an image through contrast, fluorescence microscopes use high-energy light sources of one or more narrow bands of light to excite molecules within a sample that then emit a lower energy light of a different color. The first commercial fluorescence microscopes came from Carl Zeiss and other early German microscope firms working with ultraviolet light in the early 1900s. Some of the first fluorescent signals detected through these microscopes came from naturally fluorescent molecules inside of living cells. In the early 1900s, chlorophyll and several other important biomolecules were shown to have this special property of autofluorescence. There are several images of autofluorescence in this book, including the pink cyanobacteria cells in Chapter 2.

Those early fluorescence microscopists (including August Köhler, who worked at the Zeiss company after gaining fame through his Köhler illumination method) would be happy to know that over one hundred years later, fluorescence microscopy is still an essential imaging method in the life sciences. Most fluorescence microscopy today involves, in addition to autofluorescence, the ability to selectively label individual molecules or structures that are not intrinsically fluorescent. Cells may be genetically engineered to express fluorescent proteins, often under specific conditions. We can see this in the van Gogh bundles in *Bacillus subtilis* that can appear as either red or green, depending on what the cells are doing at a given time (for example, whether the cells are producing extracellular matrix or producing surfactant during the sliding motility behavior). And fluorescent probes (also called fluorophores) are also used to stain individual cellular components within biological samples like DNA, or a specific protein via an antibody that binds to that protein, or lipids within the cell membrane. Small pieces of DNA (called oligonucleotides) that only bind to specific genetic targets are also linked with fluorescent probes. An advanced application of fluorescent DNA oligonucleotides was used to produce the image of the many different microbial species in a human dental plaque microbiome seen in Chapter 1.

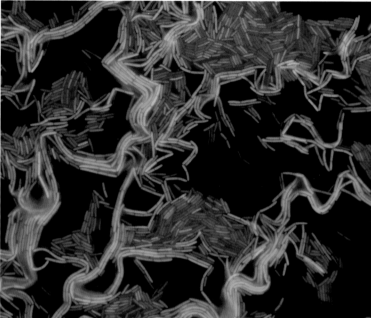

A beam of blue light excites a biofilm grown within a glass-chambered slide.

Fluorescence microscopy of van Gogh bundles formed by the bacterium *Bacillus subtilis*. Red cells are cells that produce surfactant and green cells produce extracellular matrix. The colors come from fluorescent proteins that were coupled to the promoter regions that control the expression of matrix or surfactant genes.

Confocal microscopy takes fluorescence microscopy from two dimensions to three dimensions. Confocal microscopes were conceived of in the 1950s and first built over a decade later, as documented in a paper published in *Nature* in 1969 by Paul Davidovits and M. David Egger, "Scanning Laser Microscope." The paper described a method for imaging "thick objects of low reflectivity and low optical contrast." Confocal microscopy eliminates out-of-focus light and instead scans across a sample, collecting light from one point at a time, back and forth in the x and y dimensions, and up and down in the z dimension. Then, the data collected across all points scanned are combined into one three-dimensional image or model using computer software. Confocal microscopy is now a mainstay of many fields in biology. For example, confocal microscopy can be used to tease apart individual fluorescently labeled neurons within thick tissues, such as brain slices. As applied to microbial science, confocal microscopy is equally useful for thick microbial specimens like biofilms and for other developing microbial communities where the biological information of the system is spread across space and time.

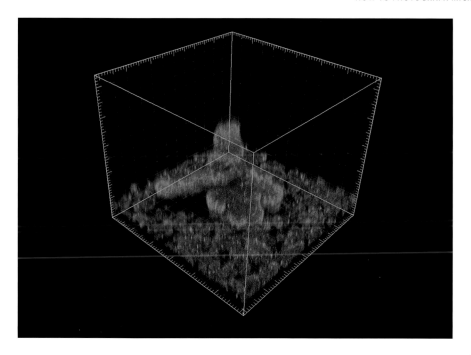

A three-dimensional model produced with confocal microscopy of an archaeal biofilm from the species *Haloferax volcanii*. The *Haloferax* cells were treated with a fluorescent dye that stains the cell membrane prior to imaging.

ENTERING THE NANOSCOPIC REALM WITH ELECTRON MICROSCOPY

Though light-based photomicroscopy has improved tremendously since the late 1830s, there are even smaller biological structures that can be visualized only with twentieth-century imaging methods completely distinct from optical systems. These subcellular or ultrastructure features are the domain of the electron microscope. The first electron microscope was unveiled by physicist Ernst Ruska and the electrical engineer Max Knoll in the early 1930s. Instead of the photon particles of visible light, electron microscopes produce images through beams of electrons. With a wavelength up to 100,000 times shorter than photons, electrons have the capacity to resolve details not dreamed of by light microscopists. They pass electrons through a subject (the premise of transmission electron microscopy, or TEM) or bounce electrons off a subject (the premise of scanning electron microscopy, or SEM). In all, electron microscopes can produce images at 10 million times magnification (10,000,000x), rather than the 2,000x theoretical limit of light microscopy.

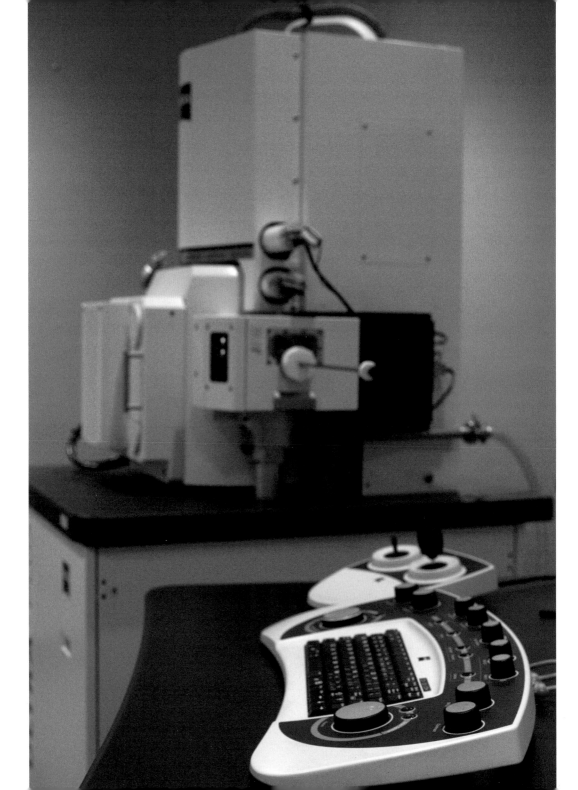

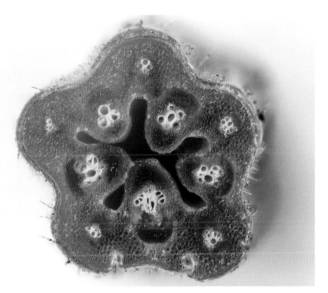

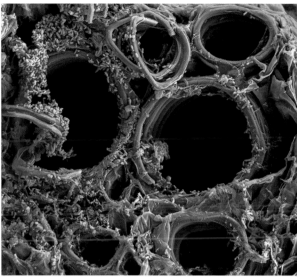

So, suppose that we examine a macro photograph of a section through a squash plant stem taken from a farm today, similar in magnification to Talbot's first prints of stems. There could be, for example, entire bacterial microcolonies of the plant pathogen *Erwinia tracheiphila* right there within the low-magnification image of the squash stem. Those *Erwinia* bacteria might affect the entire crop and the entire community that depends on that crop. And yet the individual bacterial cells and the biological materials within the colony that hold the cells together and help them to spread throughout the plant are entirely out of sight. If we take that same squash stem and prepare it for imaging on a modern scanning electron microscope, we can increase the magnification to see just about every last structure on those bacterial cells. In the resulting electron micrographs, we are looking deep inside of an individual xylem tube in the stem—a field of view that would appear as just another small circle in the low-magnification image. And in this field of view we can see the individual cells, the detail of the exterior of those cells, and the extracellular polymeric substances that hold the cells together. The downside of electron microscopy is that samples are typically fixed and often covered with a thin coating of gold or platinum alloy prior to imaging. We cannot image living cells through electron microscopy. It's a limitation that raises an important general consideration: in microscopy,

Macro photograph of a cross section through a squash plant stem.

Zooming in with a scanning electron microscope, bacteria appear within the squash stem's xylem vascular tissues. In this image, we look down ribbed tubes and see where the bacteria have congregated.

(Opposite) A modern scanning electron microscope and its controls. Rather than a single focus knob, focus and other parameters used to produce images on an electron microscope are more complicated, necessitating a keyboard control station. Almost all of the knobs seen on the keyboard are adjusted in the making of every single image collected on one of these microscopes.

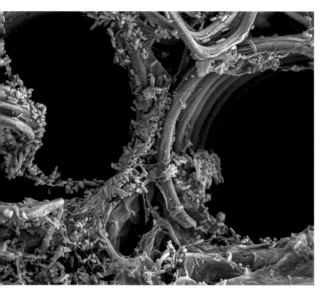

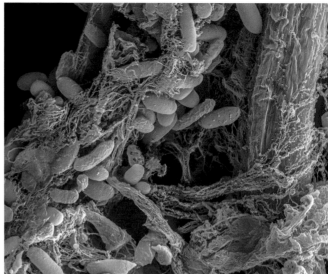

The species of bacteria inside the plant is *Erwinia tracheiphila*. Clusters of *Erwinia* cells are attached to one another through an extracellular matrix.

Individual *Erwinia* cells and their extracellular matrix seen inside the xylem. The smaller spheres seen on the rod-shaped bacterial cells are most likely outer membrane vesicles or OMVs. Outer membrane vesicles are commonly formed by gram-negative bacteria like *Erwinia* and can contain and transport DNA and other cellular cargo between cells. The magnification of this image is approximately 25,000x.

there are always trade-offs between the strengths and weakness of different imaging methods.

ADVANCED IMAGING METHODS

There are some advanced imaging methods not included in this book. Atomic force microscopy generates three-dimensional images of individual microbial cells and other extremely small structures with far greater precision than confocal microscopy. The technique uses a physical probe that literally scans over and "feels" the surface of the specimen. There are also new types of super-resolution fluorescence microscopy that have been developed in the past few decades. These use a variety of optical tricks to bypass the diffraction limitations of light, yielding an image resolution that can map single molecules. They go by a number of neat-sounding acronyms: STED (whose developers were awarded the 2014 Nobel Prize in Chemistry), PALM, and STORM. The latest entry on the scene is lattice light-sheet microscopy. Lattice light-sheet microscopes use a structured sheet of light to excite the sample. This technique causes less harmful phototoxicity in living specimens, allowing cells and tissues to be imaged at high resolution over time without being fixed (killed) beforehand or destroyed during imaging.

ANYONE CAN PHOTOGRAPH MICROBIAL LIFE

Unless you need to see individual molecules within live microbial cells or sub-cellular nanoscopic structures, you don't need a lattice light-sheet microscope or an electron microscope. You can begin your photographic journey by appreciating the macroscopic forms of microbes that are all around us. No matter where you point a camera within the natural world, you are probably photographing microbes, even if you do not realize it. There are the microbes in lichen colonies, the microbes in river rock biofilms, algae in ponds, the microbial mats in marshes and hot springs, and the microbes on foods. If you want to see individual microbes, though (except of course *Thiomargarita* and other gigantic bacterial cells), you will need a good optical microscope.

Good optical microscopes don't have to be expensive. There is now an origami microscope made of paper and a small LED light called the Foldscope that costs less than one dollar in parts. The Foldscope fits in a pocket and can survive a drop from a building, all while resolving objects at sub-micrometer resolution with 2,000x magnification. In fact, with its back-to-basics design based on a small spherical lens held near the eye, using a Foldscope is about the closest you can come to viewing the microbial world Antoni van Leeuwenhoek saw in the 1600s. For less than the cost of a sandwich, you can see more detail in the microbial world than Leeuwenhoek did. You can visualize a layer of life that no person could visualize just a few hundred years ago.

There are many ways to experience the microbial world. You can enjoy the microbial world through your senses with a bottle of wine and a slice of cheese. You can illustrate the forms and behaviors that you discover with pen and paper just as

Photographing the Grand Prismatic Spring, Yellowstone National Park, Wyoming.

Roland Thaxter, Ernst Haeckel, and many other talented scientists did, without the need for a camera. Or you can attach a cell phone camera or a more sensitive high-resolution camera to any microscope and capture in one photograph the beauty, the biology, and the mystery of microbial ecosystems. There are the microbial worlds on tiny moss leaves like the ones we encountered in Chapter 3. There are wild microbial creatures creeping like slime molds in the leaf litter of your neighborhood woods. If you dare to look, there are microbial worlds on your skin and in your mouth. You can look into this world as many times as you wish and you will never become bored by the biodiversity of microbial life.

A wild plasmodial slime mold. This slime mold was living in the leaf litter in Concord, Massachusetts, near Walden Pond.

GLOSSARY

abiogenesis   The emergence of life from nonliving or inanimate chemical precursors.

accessory genome   In the context of a population of the same species, the part of the pangenome that is not present in all strains. Also known as the variable, flexible, or dispensable genome.

actinomycin   A bioactive compound produced by *Streptomyces* bacteria that binds to DNA and in doing so inhibits RNA transcription. First isolated by Selman Waksman and H. B. Woodruff in 1940. Actinomycin D, a derivative, is used today as a chemotherapy agent in human medicine.

aerotaxis   The movement of an organism stimulated by, and in the direction of, higher oxygen concentrations.

agar growth medium   A commonly used growth medium in the laboratory containing a semisolid substance isolated from algae. Microbes on agar medium typically form macroscopic colonies as a result of continued cell division without much translocation, leading to the accumulation of large numbers of cells in a discrete location.

algae   A catch-all term of convenience for many types of photosynthetic organisms—from microbes, such as diatoms, to giant kelp. In many cases, the similarities across algal species are due not to close relatedness but rather coincidences of parallel evolution.

aminoglycoside   A class of antibiotics produced by *Streptomyces* bacteria that kill other bacteria (mostly gram-negative species) by binding to the ribosome and inhibiting protein synthesis. This class of antibiotics includes kanamycin, gentamicin, neomycin, and streptomycin.

amoeba   A diverse and broad group of protist organisms typified by movement through the extension of fingerlike projections. The group includes the genus *Amoeba*, the prototypical microscopic species *Amoeba proteus*, and macroscopic amoeboid organisms such as slime molds.

amyloid   Proteins and polypeptides that have the capacity to aggregate into sticky fibril (fibrous) structures. In microbes, amyloids are a common constituent of the biofilm extracellular matrix and a mechanism of cellular adhesion. In humans, proteins that form amyloids are often associated with diseases, such as Alzheimer's disease.

anaerobic   A state of living that does not require oxygen. An organism of this type is called an anaerobe.

animalcule   An outdated term once as widely used among scientists as *microbe* is today. Common in the work of Antoni van Leeuwenhoek and other early modern microbiologists, *animalcule*, and the idea of microscopic animals, dates back over two thousand years to the Roman author Marcus Varro. Varro speculated that animalcules multiply in swamps and other places and are then inhaled by humans.

antibiosis   An antagonistic interaction among two or more organisms, often caused by chemicals they release during the interaction.

antibiotic   A molecule known to kill or inhibit the growth of a particular group (narrow spectrum) or wide range of microbes (broad spectrum), particularly at high concentrations. Antibiotics may be chemically synthesized or produced naturally by microbes themselves. Likely to have entirely different roles within microbial ecosystems, including as signaling molecules.

antibiotic resistance   The loss of sensitivity to antibiotics that were previously effective in the killing or inhibition of a given microbial species. Antibiotic resistance arises when genetic mutations within the targeted microbe's genome, or genes acquired from other microbes, deactivate the antibiotic or alter the molecular binding site of the antibiotic. Resistance often spreads rapidly throughout microbial populations in response to widespread antibiotic treatment because the relatively few cells in a microbial population that become resistant survive exposure. As a result, they are more successful, from an evolutionary perspective, than their sensitive siblings.

Archaean eon   (4.0–2.5 billion years ago [bya]) The period when microbial life took hold and diversified across the planet.

archaellum   The archaeal analog of the bacterial flagellum. Like the bacterial flagellum, the archaellum is an appendage that rotates and thereby drives swimming motility. However, the flagellum and the archaellum evolved independently and are assembled by distinct genetic components.

arms race (evolutionary)   In the context of prolonged interaction among two or more organisms within an ecosystem, a type of co-evolution whereby an evolutionary change driven by genetic mutation in one organism is matched by another organism in response. A classic example is predator-prey interaction. As a population of predators becomes more efficient at killing prey (for example, by developing a mutation that allows it to sprint faster), the prey population becomes more efficient in evading the predator (for example, by developing a mutation that enables it to run longer distances and outlast sprinting predators).

artificial selection   A type of human-driven evolution in other species guided by deliberate reproduction of individual organisms having traits considered desirable by the selector.

astrobiology   The scientific field that unites astronomy and biology, with the main goal of looking for and identifying life on other planets. Also known as exobiology.

atmosphere   The gaseous layer surrounding Earth and other celestial bodies, held together by the body's gravitational pull. Earth's atmosphere is around 78 percent nitrogen, 21 percent oxygen, and 0.9 percent argon, with trace amounts of other gases such as carbon dioxide.

atoms   The subunits of matter that determine the physical properties of elements. These are the subunits of molecules, and molecules the subunits of life. Unfathomably tiny, they exist in a physical realm far smaller even than the microbial world, typically on the order of ten-billionths of a meter.

autofluorescence   The natural emission of light by any cell or, more specifically, individual molecules within those cells, including pigments such as chlorophyll. These molecules produce light without the intentional addition of fluorophores.

B

bacillus (plural bacilli)   A rod-shaped microbe.

bacteriovore   An organism that eats bacteria.

banded iron formation   A type of sedimentary rock comprising layers rich in iron separated by thin layers deficient in iron. Banded iron formations are primarily found in geological formations between 2.4 billion and 1.8 billion years old and were likely formed within ancient oceans when oxygen produced by photosynthetic microbes interacted with dissolved iron.

basidiospore   The reproductive spore of a basidiomycete fungus. For example, the edible mushroom *Agaricus bisporus*.

bilayer sheet   A precursor structure to a basic membrane's vesicles or cell membrane. Bilayer sheets self-assemble through the interaction of fatty acids or phospholipid micelles.

binary fission   A common mode of asexual reproduction in which one cell divides into two daughter cells. Mainly seen in bacteria and archaea but also occurring in some eukaryotic microbes.

biochemical oscillator   In the slime mold *Physarum polycephalum*, the hyper-interacting subunits within one large plasmodial cell that underpin complex developmental processes, collective behaviors, and decision making, such as cytoplasmic streaming and fruiting body formation.

biodiversity   The variety of all life forms and the variability among these. The term may be used in the context of a single microhabitat or microbial ecosystem, macroscopic ecosystems, even Earth's biosphere as a whole.

biofilm   A community of microbes living on a surface and forming a discrete structure. Biofilms comprise many cells from one or many species and an extracellular matrix holding these in place. Biofilms occur naturally on solid submerged surfaces, floating at the liquid-air interface. They also occur on solid surfaces exposed to air. Biofilms are grown and studied in the laboratory, often on agar growth medium or on submerged surfaces such as glass slides.

biogeochemical cycles   The movement, over time, of an element or compound through living and nonliving forms and places across Earth. Examples include the carbon and nitrogen cycles.

biological entity   A broad label for anything with some or all of the properties of life. Cellular organisms and viruses are biological entities. Early forms of life on Earth and hypothetical alien forms of life on other planets, which may be different from cellular organisms as we understand them today, are also biological entities.

biomass   A way of enumerating or estimating the living fraction of any ecosystem by mass. For example, if the microbial cells and other organisms were filtered or otherwise captured from the ocean, separating the free water from all life, the total mass of that living fraction would be the biomass of the ocean.

biome   A community or ecosystem of organisms inhabiting a particular geographic location or environment. The composition and dynamics of organisms in a given biome are determined by physical conditions of the environment and are therefore different from one biome to the next. In the macroscopic realm, deciduous forest is one example of a biome. In the microbial realm, the human body itself is a microbiome, containing a multitude of distinct biomes.

biosignature   A sign of past or present life, on Earth or an extraterrestrial body.

biosphere   The totality of organisms and ecosystems on Earth, from the microbes in the deepest sediment below the ocean floor, to those surviving high in the atmosphere, and all in between.

Black Queen hypothesis   An evolutionary hypothesis that explains how one species within a mixed-species population can lose a gene over time if the product of that gene is provided by another species that leaks that same product into the environment. By eliminating the gene, the first species saves the energy cost of generating the product, while still enjoying the benefits of the product thanks to the interacting species.

built environment   The environments constructed by, lived in, and worked in by humans. The built environment is a relatively new habitat.

C

calcareous ooze   A white, chalky mud found on the ocean floor, composed largely of the calcium carbonate exoskeletons (coccoliths) of coccolithophores. This calcareous ooze is compacted over time and eventually forms the sedimentary rock we know as chalk.

Cambrian period   (541–485 million years ago [mya]) Includes the Cambrian explosion ~541 mya, when, according to the geological record, many animal groups appeared.

capsid   The outer shell-like structure of a virus, made of proteins. For many bacteriophage and archaeaphage, the capsid looks like the head of the virus.

carbon fixation   The process of converting inorganic carbon in the form of carbon dioxide gas (CO_2) to organic carbon compounds, as carried out by living organisms. Also called carbon assimilation.

carotenoid   A class of yellow, orange, or red pigments that occurs naturally in many different types of organisms, as seen in cyanobacteria living in microbial mats in Yellowstone National Park and in the haloarchaea of Great Salt Lake.

cell   The smallest functional and structural unit of biological organization in almost all organisms on Earth. Some organisms force slight relaxation of this basic definition while remaining fundamentally cellular. For example, plasmodial slime molds reach a large macroscopic size but comprise one contiguous, multinucleated cell. Their complex behaviors are the result of interactions among subcellular biological units.

cell biology   The scientific study of cells and, in particular, the ultrastructural features within cells. Cell biology was once entirely concerned with eukaryotic cells;

however, the field has expanded as it has become clear over the past few decades that bacteria and archaea have cytoskeletons and other facets of complex subcellular organization.

cell fusion   The physical joining of two cells, either transiently or for longer durations. Usually these will be cells from the same species, as in the cellular-rejuvenation behavior of the bacterium *Myxococcus xanthus*. Plasmodial slime mold cells also may fuse.

cell rejuvenation   A cooperative behavior among a population of interacting microbial cells during which cells fuse with one another and are able to exchange sections of the cell membrane. As seen in the bacterium *Myxococcus xanthus*.

central dogma   A concept (in some ways oversimplified) describing the flow and processing of genetic information in cellular organisms: DNA encodes genetic information in units called genes; these are transcribed into messenger RNA intermediates; and these RNAs are translated into proteins through the action of the ribosome along with transfer RNAs.

chalk   A soft and porous sedimentary carbonate rock composed of calcium carbonate, also known as calcite (a form of limestone). Formed as the calcite discs of coccolithophores settle and accumulate on the ocean floor.

chlorophyll   Several related green pigments seen in cyanobacteria, algae, and green plants that enable photosynthesis through the absorption of light energy and conversion of that light energy into chemical energy within the cell.

chloroplast   The organelle within green plant and algal cells that carries out photosynthesis. Derived from an ancient endosymbiosis between a cyanobacterium and a eukaryotic cell already containing mitochondria.

chromosome   A structural and organizational unit of an organism's genome. This complex of DNA and proteins is held inside the cell. Some organisms have many chromosomes; others, including many bacteria and archaea, have one or a few chromosomes in addition to smaller genetic elements such as mini-chromosomes and plasmids.

cilium (plural cilia)   An organelle in eukaryotic cells that extends from the cell either. Cilia may serve as environmental sensors or as appendages that beat rapidly, driving motility or feeding.

coccolith   A calcium carbonate disc produced inside of coccolithophores, expelled from and attached to the outside of the cell.

coccosphere   The spherical exoskeleton of a coccolithophore, composed of many coccoliths.

coccus (plural cocci)   A circular or spherical microbe, sometimes arranged into groups that look like clusters of grapes or long chains.

collective behavior   Apparently organized phenomena lacking any leader or otherwise centralized direction. Instead, the outcome is a result of group dynamics and local interactions among individual entities.

colony   A collective form of several or great numbers of organisms living together in close physical proximity. Typically the term describes collective forms of organisms derived from a single parent cell, as in the macroscopic formations of microbes on agar growth medium or in natural settings.

competition   Biological interaction between two or more species or between different individuals of the same species vying for limited nutrients or other resources. At play in the microbial world as much as in macroscopic environments.

composite organism   A cohesive living entity that looks like an organism but is inconsistent with the traditional definition of an organism, such as lichens. The inconsistency arises because composite organisms comprise at least two distantly related species and have properties the independent organisms lack.

confocal microscopy   An advanced form of fluorescence microscopy yielding increased resolution and contrast and used to produce 3D models of microscopic structures.

conidia   Fungal spores produced and released by hyphae.

conidiophore   A specialized structure with a bouquet-like appearance occurring on fungal hyphae, which generates and releases conidia.

conjugation   Horizontal gene transfer as mediated by a physical structure bridging two cells.

conservation of biology   The idea that biological processes such as the chemical steps leading to the origin of life are still occurring somewhere on Earth today.

cooperation   The collective effort of a group of organisms resulting in mutual benefit. Cooperation often occurs among individuals of nearly identical or closely related organisms, but also occurs between distantly related species. Cooperation between distantly related species is called mutualistic symbiosis. The importance of cooperation is obvious at the macroscopic level; it is just as relevant, albeit invisible, among microbes. The idea of extending cooperation to microbes dates back to the early twentieth-century writings of Russian polymath Peter Kropotkin.

core genome   Within a population of the same species, the part of the pangenome present in all strains. Includes genes that are most essential to the basic functions of the cell.

cosmic microwave background   Radiation lingering from a few hundred years after the beginning of the universe as we know it (the Big Bang). Detectable by advanced space-based telescopes, this radiation has been used to determine the age and evolution of the universe.

cosmic ray   High-energy radiation that originates from the Sun and from supernovae and other sources outside our Solar System. Cosmic rays travel through the universe at nearly the speed of light. They are damaging to life and electronics, but most are blocked by the magnetosphere of Earth and Earth's atmosphere.

cosmic web   The largest scale structure in the universe, a distribution of galaxies into galaxy clusters and galaxy filaments and galaxy walls determined by dark matter.

cosmos   The universe regarded as a complex and yet orderly system. The scientific discipline concerned with the origin, current state, and future trajectory of the cosmos is known as cosmology.

Cretaceous period   (145–66 mya) The period ending in a mass extinction known as the K-Pg boundary. Most large animals, including the dinosaurs, died.

cryophile   A cold-loving organism. Also known as psychrophiles, these live at temperatures considered extremely low according to the standards of other forms of life.

cryptobiosis   A state of inactivity assumed by tardigrades, rotifers, and other micro-animals in response to dry conditions. The cryptobiotic forms of these creatures are so inactive that metabolism and other signs of life are virtually undetectable.

cytoplasm   The aqueous interior of cells. A dense gel-like milieu containing the organelles and all components needed for cellular metabolism.

cytoplasmic streaming   A pulsing behavior of slime molds that allows them to move, sense their environment, and engage in complex behaviors.

cytoskeleton   The internal structure of a cell, composed of structural filaments. Bacterial and archaeal cytoskeletons contain some elements related to eukaryotic filaments. The cytoskeleton affects cell division, shape, polarity, and other functions and features of cells.

cytotoxic   Anything toxic to living cells. Typically used in reference to human or other animal cells.

D

dark matter   An unidentified type of matter that makes up a major portion of the universe. Dark matter does not emit or interact with electromagnetic radiation, but it can be observed indirectly through its effects on massive assemblages of ordinary matter, such as galaxies and galaxy clusters.

Darwinian evolution   The change in characteristics of any population of biological entities occurring through transmission of heritable material (i.e., genes) through successive generations. Heritable traits arise due to natural variation in the genetic code. The theory of natural selection, as originally proposed by Charles Darwin, holds that if these traits improve the survival of an organism and increase its reproductive capacity, they will become more prevalent in future generations of that organism.

deep biosphere   A vast microbial ecosystem existing in complete darkness within subsurface sediments across the Earth, including kilometers deep below the ocean floor. This unseen biosphere makes up a major fraction of the planet's biosphere.

dental plaque   A biofilm that develops and persists on the surface of teeth. Most microbial species in the human mouth microbiome do not cause harm, but acid-producing microbes such as the bacterium *Streptococcus mutans* do spur the breakdown of the enamel matrix and cause dental caries ("cavities").

diffraction limit (of visible light)   The size limit of a small object or distant large object past which no further detail can be resolved with a microscope or telescope. Determined by the physical properties of light, primarily the wavelength of visible light.

division of labor   An advanced form of cooperation seen in social groups and in multicellular organisms, whereby individual cells or organisms are divided into different functional and synergistic groups that carry out different tasks.

domain   See the Domains of Life list at the end of the Glossary.

domestication   A relationship between two groups of organisms over many generations in which each benefits and one significantly influences the evolution of the other, typically through the selective reproduction of some individuals over others (artificial selection). Examples include domestication of dogs, plants, fungi, and many other species by humans and domestication of fungi by leaf-cutter ants.

DNA   Deoxyribonucleic acid. The biomolecule, shaped as a double-helix of two strands composed of nucleotide subunits, that encodes genetic information and is the heritable material in all cellular organisms and almost all entities on Earth, except RNA viruses.

E

Earth's core   The inner region of the planet, with a radius of about 3,400 kilometers. Composed mostly of iron and nickel and divided into a molten outer core and dense inner core of solid mass or plasma as dense as a solid.

Earth's crust   The outermost region of the planet comprising a mixture of metamorphic, sedimentary, and igneous rocks. It is divided into oceanic crust and continental crust and varies from five kilometers to fifty kilometers in thickness.

Earth's lithosphere   The outer region of Earth including the crust and the upper mantle. These two layers act as a single mechanical layer in plate tectonics.

Earth's mantle   The region of the planet sandwiched between the crust at the surface and the outer core. Accounts for around two thirds of Earth's mass and three quarters of Earth's volume. Appears solid at any given moment but is actually a viscous fluid over geological time.

ecological succession   A change in an ecosystem over time. Each stage in the succession features a different complement of species, determined by the activities of species in preceding stages.

ecosystem   A community of organisms interacting with each other and with the non-living components of the environment.

ectosymbiont   An organism or biological entity that attaches to the exterior of another with which it lives in symbiosis. In contrast to endosymbionts, which live in symbiosis inside of another cell.

edge of sight   A hypothetical limit for any organism with visual organs, beyond which objects cannot be resolved with the naked eye. Determined by visual acuity.

efflux   The movement of chemicals or other substances out of a cell.

electron   An exceedingly small and light subatomic particle with a negative charge necessary to the physical phenomena of electricity, magnetic fields, conductivity, and heat. Thought of as an elementary particle because we do not know of any substructure that composes it.

electron microscope   Any microscope that creates images using a beam of electrons rather than light to resolve microscopic details. The two main subtypes are transmission electron microscopy and scanning electron microscopy.

emergent property   Any property of a system composed of many distinct subunit parts that cannot be accomplished by the subparts alone. A synergistic phenomenon.

endosymbiosis   The process of one cell being taken up by and living inside of another cell permanently. The cell taken up is called an endosymbiont.

enzyme   A protein or RNA molecule that catalyzes one or more chemical reactions. In the context of these chemical reactions, the molecules that enzymes act upon are known as substrates.

eumicrobe   An organism or biological entity that can never be seen with the naked eye. Every aspect of a eumicrobe's natural lifestyle is microscopic. All viruses are eumicrobes. Some bacteria and archaea, but not all, are eumicrobes. The term eumicrobe employs the Greek prefix *eu*, meaning "good" or "real." Can also be thought of as "obligate microbes."

eusocial   A special designation for organisms that engage in certain kinds of advanced social organization and behavior, including many species of bees, ants, wasps, and termites; some crustaceans, such as snapping shrimp of the genus *Synalpheus*; and some mammals, such as the naked mole rat. E. O. Wilson, who coined the term, argues that humans are eusocial, but the matter is controversial.

exoenzyme   An enzyme secreted by a cell into the extracellular space, often to break down nearby polymers into small pieces to be taken up by the cell.

exoplanet   Any planet that does not orbit our Sun. Also called an extrasolar planet.

external memory   The type of navigational memory used by *Physarum* and other slime molds, based on extracellular slime trails left behind by the plasmodial network as a surface is explored.

extracellular DNA   DNA molecules existing outside of cells. These are ubiquitous in natural environments. They are crucial in the ecology and evolution of organisms and in the biochemical cycling of several of life's most critical elements, including phosphorus. Also called free DNA or naked DNA.

extracellular matrix   The noncellular component of biofilms that holds the cells together and carries out several important biological functions. A mixture of many different types of extracellular polymeric substances, including exopolysaccharides, extracellular DNA, lipids, vesicles, amyloids, enzymes, and other proteins in an aqueous environment.

Extreme Deep Field   A composite image stitched together from many long-exposure images collected by the Hubble Space Telescope, revealing the distant galaxies in a tiny patch of the night sky. These objects are far too dim to see with the naked eye.

extremophile   A microbe that lives normally under extreme environmental conditions, including severe hot and cold, high or low pressure, high salinity, or high acidity/alkalinity. This term is convenient, but it does not carry special biological significance. After all, extremity is just measured relative to conditions preferred by humans and other animals we know well.

F

facultative multicellularity   The condition in which the multicellular form may be relevant to the ecology and evolution of the organism but is not required for reproduction and completion of the life cycle.

farming   The application of agriculture: a relationship between two or more organisms in which one cultivates and drives the reproduction of the other(s) for sustenance or other beneficial uses.

fatty acid   A form of carboxylic acid—a small, naturally occurring chemical—with a chain of carbon atoms that can self-assemble into micelles and through further steps into membrane vesicles.

federation   An assembly of many subpopulations of cells, each adapted to a particular ecological niche, but together identifiable as one species due to a common core genome. The localized adaptations are enabled by a diverse accessory genome that varies across the entire federation. Also called metapopulation.

fermentation   A microbial decomposition process yielding acids, alcohol, and other desirable microbial products and flavors. In the strict sense, microbial metabolism when there is no cellular respiration and ATP is wholly derived from using organic compounds as terminal electron acceptors.

fitness   The suitability of an organism to survive and reproduce in a given environment. Fitness is determined by the organism's genetic traits (genotype), manifest in physical traits (phenotype).

flagellum (plural flagella)   A cellular appendage used by bacteria for swimming locomotion. From the Latin meaning "whip."

fluid inclusion   In reference to halite salt crystals, a tiny pocket within the crystal where liquid water and potentially halophilic microbes are trapped and secluded from the external environment.

fluorescence   The physical phenomenon whereby a substance absorbs light of a particular wavelength and, as a result, emits light, typically of longer wavelength and lower energy.

fluorescence microscopy   A method of optical microscopy based the detection of fluorescence rather than, or in combination with, absorption and reflection of light.

fluorophore   A chemical that fluoresces, often used to label biological structures either directly or indirectly when conjugated to another molecule.

food web   The network of ecological connections between organisms in an ecosystem. Often shown graphically as a who-eats-whom diagram of the intersecting food chains of many species.

free-living   The condition of an organism that can reproduce on its own.

fruiting body   A multicellular structure produced by fungi, slime molds, and some bacteria that produces and disperses reproductive spores. Usually macroscopic.

G

galaxy   A system of stars bound together by gravity along with intervening remnants of stars, interstellar gas, dust, and dark matter.

galaxy cluster   Cosmic structures of galaxies bound by gravity.

galaxy filament   These threads of galaxy clusters, typically hundreds of millions of light-years in length, are the largest structures observed in the universe. Between galaxy filaments are cosmic voids—areas containing few or no galaxies.

gamma ray   Electromagnetic radiation produced when high-energy photons are released by atoms undergoing radioactive decay. An ionizing radiation damaging to living organisms.

gene   A region (locus) of DNA and the basic unit of genetic information for life on Earth. Often, one gene contains the information needed to produce one protein.

generation time   The time between the birth or division of one organism (the parent) and the birth or division of its offspring.

genetic information   The information held in any heritable molecule, such as the sequence of nucleotides in DNA, that replicates and can be copied into the next generation and transferred between genomes through horizontal gene transfer.

genome   The sum of all genes found within a particular organism, organized into one or more chromosomes and other replicating genetic elements (replicons).

geological history   A time scale based not on human history but on the geological processes of Earth, where major changes occur over time periods not directly perceivable by humans.

geosmin   A compound that gives dirt its characteristic odor.

geothermal feature   A broad term for an object in the environment formed by geothermal activity in a particular region, often where magma nears Earth's surface or along fault lines of tectonic plates along the ocean floor. Includes hot springs, geysers, fumaroles, mud pots, mud volcanos, and hydrothermal vents.

germ   A misleading term commonly used to refer to microbes, as if all were dangerous contaminants. These assumptions are false. Most microbes do not cause harm; they are a normal and unavoidable component of all environments.

giant bacteria   Any bacterial species with individual cells large enough to be seen by the naked eye, including *Epulopiscium fishelsoni* and *Thiomargarita namibiensis*.

gongylidia   Swollen structures that develop on fungi farmed by leaf-cutter ants.

gram-negative and gram-positive bacteria   A division between groups of bacteria based on a laboratory staining technique with the dye crystal violet, known as the Gram stain. The technique makes individual bacteria easier to see and differentiates bacteria with thick cell walls made of the polymer peptidoglycan (gram positive) from those with a thin peptidoglycan cell wall (gram negative). The distinction is useful as a "quick and dirty" way of classifying bacteria because, for the most part, gram-negative species are more closely related to other gram-negative species than they are to gram-positive species and vice versa, giving some clues as to the identity of species under investigation.

grazer   An organism that consumes another living organism but, unlike a true predator, does not kill that organism. Often the organism consumed is a continuously growing plant or other multicellular organism. Examples include mites that eat fungi on a cheese rind and buffalo that eat grass.

gut-brain axis   A reference to the neurological effects of chemicals produced by the robust microbial communities living in the intestines of humans and other animals.

H

Hadean eon   (4.6–4.0 bya) The formation of the planet. Between 4.2 and 4.0 bya, chemical evolution may have led to early forms of life and to the last universal common ancestor.

halophilic   A salt-loving organism. Halophiles live optimally at salt concentrations from around 10 percent to above 20 percent, a condition that would kill most other forms of life. These can be differentiated from halotolerant organisms, which live at concentrations greater than seawater (around 3.5 percent in the modern oceans) but not "extreme" levels.

haploid   A state in which eukaryotic cells contain just one of the two sets of chromosomes needed for the organism's complete life cycle. In animals, sperm and ova (gametes) are haploid.

heavy bombardment   A period of frequent and severe asteroid impacts during Earth's early history. Geologists and planetary scientists agree that there was such a period, but its exact timing and duration, and the possible occurrence of a "late heavy bombardment" period, are controversial.

heterotrophs   Organisms that cannot fix carbon from inorganic sources but rather use organic carbon sources for growth, as opposed to autotrophs, which can fix inorganic carbon (CO_2) via photosynthesis.

holobiont   An ecological unit or biological entity assembled from many different physically associated organisms (bionts). A coral, with its many microbes, may be considered a holobiont. Given its complex and dynamic microbiome, the human body may also be counted as a holobiont.

hologenome   The collection of all the genomes present in a holobiont.

horizontal gene transfer   When genes are transferred between the genomes of different organisms or biological entities (horizontally) rather than from parent to offspring (vertically). A mechanism for evolutionary change, horizontal gene transfer has had major impact on the web of life.

hot spring   A geothermal feature produced as heated groundwater rises through Earth's crust.

housekeeping gene   Genes required for cellular maintenance and basic cellular function. They are typically expressed all the time (constitutively) within one or more cells of an organism.

human microbiome   The totality of the thousands of microscopic species from all three domains of life that live on and within the human body, together amounting to trillions of microbial cells.

hydrothermal vent   A geothermal feature caused by hot water rising from a fissure in Earth's crust, often seen as white or black smokers at the apex of two tectonic plates along the ocean floor.

hyper-swimmer   In microbial species, a type of cell covered by many flagella and capable of rapid locomotion, usually alongside many other cells moving together as a multicellular raft.

hypha (plural hyphae)   A branched filamentous structure seen in many fungi, actinobacteria, and other organisms. Thick bundles of hyphae are called mycelia.

I

inhibition zone   The physical area around a microbial community (e.g., a biofilm or colony) where other organisms cannot grow due to the secretion of antimicrobial molecules.

inoculation   The intentional addition of one or more microbial species to something, such as a sterile growth medium.

integrative science   Any systematic study of physical and natural phenomena that draws from multiple traditional scientific disciplines and subdisciplines. Also called interdisciplinary science.

intercalation   In molecular biology, the process of one molecule inserting into the structure of another, such as DNA.

ionizing radiation   Any type of high-energy radiation, such as gamma-rays, X-rays, and far ultraviolet radiation, capable of freeing electrons from atoms. Cells are damaged when biomolecules are ionized by radiation.

iron catastrophe   A hypothesized event in Earth's early history, when the planet was still molten hot and its major layers initially formed. During this event, dense elements, including iron, collected in the area now called the core, and lighter elements remained at or near the surface.

island universe   A previous understanding of the observable universe in which the Milky Way galaxy was thought to be the entirety of the universe.

J

Jurassic period   (201–145 mya) A period dominated, at least macroscopically, by dinosaurs and featuring great biodiversity in large marine and terrestrial reptiles.

K

kin   Organisms that are closely related to each other or share identical or nearly identical genomes.

kin discrimination   The ability of an organism to respond differently to, or cause differential effects in, organisms that are not close genetic relatives. A mechanism of cooperation and competition seen in animals and extended to similar phenomena observed in the microbial world.

L

last universal common ancestor (LUCA)   The hypothetical organism that existed at some point soon after the origin of life and became the ancestor of all cellular life on Earth today.

leghemoglobin   The oxygen-binding protein found in root nodules.

lichen   A composite organism containing one or more photobionts (either a cyanobacterium, an alga, or both) and a mycobiont fungus.

life   A complex system of chemical entities containing all functional features necessary for reproduction and change through Darwinian evolution. Living things have many emergent properties not seen among purely chemical entities, including growth and reproduction, synthesis and breakdown of chemicals (metabolism), higher-order biological organization (including cells, organisms, and ecosystems), and physiological and behavioral response to environmental conditions, including maintenance of a constant internal state (homeostasis).

life cycle   The series of biological forms required for the reproduction of a given organism.

light   In the broadest scientific sense, radiation spanning the electromagnetic spectrum. Listed from shorter wavelengths to longer: gamma rays, X-rays, ultraviolet, visible light, infrared, microwaves, and radio waves.

light microscope   Any microscope that resolves objects with visible light and magnifying lenses. Encompasses many varieties and methods, including bright-field and dark-field microscopy, differential interference contrast, and phase contrast. Also called an optical microscope.

light organ   An animal organ containing a symbiotic community of bioluminescent bacteria. Examples include Hawaiian bobtail squid and flashlight fish.

living system   A broad term for any biological entity or system of interacting biological entities. Smaller-scale living systems (i.e., individual organisms) are nested within larger living systems; physically larger systems are more complex. An interacting population of unicellular bacteria, microbial biofilm, biomes, and human bodies are all examples of living systems. The largest and most complex conceivable living system is the entire biosphere of Earth

locomotion   The many different ways of moving through the environment. In the case of microbes, these include solitary and collective forms of swimming driven by flagella (e.g., swarming), twitching (driven by pili), and sliding (driven by extracellular matrix and colony expansion).

luminescence   A broad term for light emitted by a chemical at low temperatures, including bioluminescence.

lux gene operon   A sequence of genes in bioluminescent bacteria encoding components that work together to produce luminescence through the action of the enzyme luciferase.

M

macroscopic   Visible to the naked eye.

magnetosphere   A magnetic field generated within Earth's core, extending beyond the atmosphere into space. The magnetosphere shields Earth from solar wind. This phenomenon occurs on other astronomical bodies as well.

magnification   The increased size of an object's image compared to its actual size. For example, if a bacterium is one micrometer long and its image is one thousand micrometers (one millimeter) long, the image is magnified one thousand times.

metabolism   The chemical reactions and processes that synthesize (anabolism) or break down (catabolism) molecules within living organisms.

metagenomics   The scientific study of genetic information extracted from a particular environment.

metaorganism   A population or community of interacting and interdependent organisms.

methanogen   A microbial species that produces methane. Methanogens do not form a cohesive evolutionary group, but all are within the domain Archaea.

micelle   A spherical aggregate that self-assembles from amphipathic molecules (i.e., those with hydrophobic and hydrophilic properties). An example is fatty acids, which assemble into bilayer sheets.

microaerophile   A microbe that uses oxygen but must live in habitats with low oxygen concentrations relative to the atmosphere, which contains 21 percent oxygen.

microbe   An organism or biological entity too small to be seen with the human eye. Microbes can either be sometimes microscopic in nature (facultative microbes) or always microscopic (eumicrobes).

microbes first   The theory holding that every time life has evolved or will evolve from chemical precursors within the known universe, it will begin microscopically.

microbial cloud   The microbes of the human microbiome that emanate from the body and colonize built environments, such as homes and transportation systems.

microbial community   A community of microbial species nested within larger ecosystems, possible located in or on larger organisms. Also called a consortium.

microbialite   A broad term for sedimentary structures formed by interaction with microbes, including layered stromatolites and unlayered thrombolites.

microbial macroscopic manifestation   Any manifestation of individually microscopic cells that can be seen with the naked eye.

microbial mat   A macroscopic multispecies community of syntrophic microbial organisms arranged within a layered or laminated structure.

microbial sciences   Integrative sciences drawing from all scientific disciplines that intersect with microbial life either directly or indirectly.

microbial world   A poetic way of thinking about Earth's microbial ecosystems. A term expressing the idea that the differences in scale between most microbes and the human body and senses are so great that we are surrounded by an invisible biological world that nonetheless profoundly affects all that we can see.

microbiome   A microscopic ecosystem, nested within larger macroscopic ecosystems.

microbiosphere   A conceptual subdivision of the biosphere, differentiating microbial and macroscopic life, and the early Earth biosphere, which contained only microbial life. That early microbiosphere led to microbial macroscopic forms, including stromatolites, and then to the evolution of eukaryotes and other complex organisms. Despite the evolution of larger organisms, the microbiosphere is never eliminated. Rather, it remains foundational; to eliminate it would be to pull the rug out from beneath the global ecosystem.

microcolony   A cluster of microbial cells consisting of hundreds or thousands of cells rather than millions or more. An early stage in the biofilm life cycle.

microenvironment   A microscopic area whose physical conditions differ from those of other nearby locations. Also called a microhabitat.

microscopic scale   The size scale including physical entities measured in micrometers, or millionths of meters. The domain of microorganisms and of plant and animal cells.

microwave   Electromagnetic radiation with wavelengths longer than visible infrared and shorter than radio waves, in the range of one millimeter to 0.3 meters.

migration   Collective movement of organisms (usually of the same species) in the same direction.

mitochondria   Derived from endosymbiosis of a bacterium, these organelles are found in nearly all eukaryotic cells. (Some exceptions, such as red blood cells, are derived from cells containing mitochondria. In some cases, species previously possessing mitochondria have lost them.) Mitochondria are associated with energy generation. They also produce key cellular signals and are involved in cell-cycle regulation and cell death. They vary in size across species and are usually between one and three micrometers in diameter.

molecular spear   A cellular structure injected by one cell into another, in a potentially deadly attack. Often used to describe the type VI secretion system, a particular mechanism of bacterial competition that delivers toxic cargo molecules.

morphology   The shape of a cell or larger cellular structure. Microbial cells come in many shapes including rods, cocci, filaments, bulbs, stalks, spindles, and spirals.

mostly microbial   The theory holding that most biodiversity on a living planet remains microbial throughout the entire duration of life on that planet. The mostly microbial state results from the legacy of the microbial origin of life (the microbes-first theory), the evolutionary trajectories set forth by the initial microbiosphere, and the conditioning of the planet by microbes.

motility   The ability of an organism to move through its environment. In microbiology, often refers to solitary cellular motility as driven by flagella or archaealla.

multicellular organism   Any organism composed of more than one cell. Traditionally thought of as cells united within one physical entity, however multicellular behaviors can also occur among populations of nearby disconnected cells.

mushroom   The visible and sometimes edible fruiting bodies of fungal species.

mutation   A permanent change in the sequence of any genetic element. A major source of genetic variation acted upon by natural selection in the process of Darwinian evolution.

mutualism   A type of symbiosis in which each of the partners benefits.

mycelia   Bundles of filamentous hyphae, as seen in fungi.

mycorrhiza   Fungi in symbiotic association with plant roots.

N

nanoscale or nanoscopic   The size scale including physical entities measured in nanometers, or billionths of meters. The domain of subcellular structures, viruses, and biomolecules.

natural product   Any molecule synthesized by (in the strict sense), or derived in any way from (a broad definition), living organisms. Also called small-molecule natural products.

nature   In the broadest sense, the fabric of physical reality as we know it. In daily use, synonymous with the natural world, meaning the living world on Earth.

nematocyst   A stinging cell seen in corals and other Cnidarians. Used for defense or to catch prey.

nested endosymbiosis   The evolutionary history of any organism containing one or more endosymbionts that themselves were a product of endosymbiosis. Akin to Russian dolls, with each fit into the next largest.

neurotransmitter   A chemical that acts as a signal between neurons, transferred through synapses.

niche   A localized part of a microscopic or macroscopic environment with unique physical and chemical properties that favor the growth and well-being of a particular organism or collection of organisms.

nitrogen fixation   The process of converting nitrogen gas (N_2), which is largely inert, into ammonia (NH_3), a form that most organisms can more readily use. A relatively small number of nitrogen-fixing bacterial species carry out this process.

nucleotide   The subunits of the nucleic acid biomolecules DNA and RNA.

nucleus   A membrane-bound organelle within a eukaryotic cell, containing its genome. Some organisms (e.g., *Paramecium*) have a distinct macronucleus and micronucleus, and only the DNA in the micronucleus is passed on during reproduction.

O

obligate multicellularity   The in which multicellularity is required for an organism to reproduce and complete its lifecycle.

observable microverse   The biology and other structures and happenings of microscopic realms, including the microbial world, as humans perceive them using microscopes and other technologies. Akin to the concept of an observable universe, the observable microverse has widened in scope as a result of modern science, beginning with the discovery of microbes, chemicals, atoms, and subatomic particles and continuing to this day with higher-resolution views of these entities.

observable universe   The universe as observed by humans through our senses and with the aid of imaging technologies. The observable universe has widened in scope over time, especially during the past few centuries as a result of modern physics and astronomy. But even if we could see absolutely everything within the universe, there could be a multiverse containing other universes that will never be detectable.

oligonucleotide   A short piece of DNA or RNA.

opportunistic pathogen   A microbe that does not always cause disease but which can under certain conditions.

optical aberration   A distortion or other optical defect within an image, particularly images produced through glass lenses.

oral groove   The mouth-like feeding structure of a paramecium.

organelle   A compartmental structure of cells that carries out one or more functions. Some of the most important organelles contain their own DNA because they were derived through the process of endosymbiosis. A stricter definition might include only these.

organism   An individual living system. Generally used to refer to cellular life on Earth. May also refer to hypothetical biological entities on other planets.

outer membrane vesicle   Common nanoscopic vesicles released by gram-negative bacteria that often traffic signals, genetic information, proteins, toxins, and other molecules between cells within a population.

ozone layer   The portion of Earth's atmosphere that contains the molecule ozone (O_3) and absorbs ultraviolet radiation from the Sun.

P

Paleolithic period   (~3 mya–10,000 years ago) A period of human prehistory when human technology was based on stone tools. Also called the Stone Age.

pangenome   The entire set of genes spread across all strains within a population of the same species, encompassing the core genome and all accessory genes present.

parasitism   A symbiosis between two or more organisms in which one organism is harmed while the other benefits. Pathogenesis is thus a form of parasitism.

pathogen   Microbial species that cause disease, including in humans. Although a convenient term, *pathogen* has little significance from a strictly biological perspective because the microbe in question is pathogenic only from the point of view of the negatively affected organism. Critics of the term prefer to think of pathogens as symbionts capable of parasitism.

penicillin   An antibiotic that prevents many bacterial species from growing and dividing by inhibiting cross-linking of peptidoglycan within the cell wall. Produced by *Penicillium* fungi and discovered by Alexander Fleming in 1928.

petri dish   The small, circular dish invented by German microbiologist Julius Richard Petri, commonly used to hold agar growth medium for routine microbiology experiments.

phage   The viruses that prey on bacteria and archaea. Short for bacteriophage or archaeaphage.

photon   The elementary particle that makes up light, first theorized by Albert Einstein.

photosynthesis   A process carried out by plants, algae, and cyanobacteria that generates chemical energy from light energy. Organisms that use photosynthesis are called photoautotrophs or simply phototrophs.

phylogenetic   Referring to the patterns of evolutionary history of organisms and the familial relationships among organisms.

phytoplankton   Photosynthetic plankton, as opposed to animal-based zooplankton. The smallest phytoplankton is called picophytoplankton.

pigment   Naturally colored biological molecules. The color of a pigment is the result of the selective absorption of some wavelengths of light and reflection of others. For example, chlorophyll pigments absorb blue and red light and reflect green light, making plants and algae green.

pili (singular pilus)   Thin proteinaceous appendages on bacterial and archaeal cells with many different functions.

planetesimal   A small planet. Also refers to the small pieces of matter that coalesce through gravity to form planets.

plankton   A diverse group of small aquatic organisms. May have the capacity for locomotion over small distances, but movement is superseded by currents within the water column. Divided into zooplankton and phytoplankton. The word is derived from the Greek *planktos*, meaning wanderer.

plasmodium   An active-feeding form of some slime molds that moves through cytoplasmic streaming.

plate tectonics   The phenomenon of continental drift and gradual reshaping of the Earth's surface over geological time driven by the movement of seven major and several smaller "plates" of Earth's lithosphere.

predator-prey interaction   Whenever one living organism consumes another, killing that organism and assimilating atoms from that organism into its own cells.

primordial heat   Energy within Earth's core lingering from the formation of the planet.

protein   Biomolecules composed of folded chains of amino acids. The chains are said to have four levels of structure. Primary structure is the sequence of amino acids. Secondary structure arises from the initial interactions of the chain's backbone leading to the formation of helices or sheets. Higher-level interactions among the amino acid side chains cause the full protein to fold into its tertiary structure. Folded chains interact with each other, constituting proteins' quaternary structure.

protist   A catch-all name for the many eukaryotic microbial species excluding fungi, micro-algae, and micro-animals.

protobiology   A hypothetical version of biology on Earth that preceded and may be different from the universal cell-based biology of organisms today.

protocell   A hypothetical cell precursor that existed before the last universal common ancestor of life on Earth, or a basic cell precursor produced from prebiotic chemicals in a laboratory.

protoplanetary disc   The region within a solar nebula where dust and gas accumulate and condense to form planets and solar systems.

public good   In ecological and evolutionary theory, a molecule produced by a cell and released into the extracellular environment, where it may used by other cells.

Q

quorum sensing   Chemical communication systems present in bacteria and archaea. Quorum sensing enables a population of microbes to engage in coordinated, collective behaviors by responding simultaneously to an accumulation of small molecules, called autoinducers, in the environment. Allows microbes to indirectly sense which other microbes in the area are from the same species (intraspecific) and which are not (interspecific).

R

radioactive decay   The natural phenomenon whereby the nucleus of an atom randomly loses energy as it emits radiation. The radiation emitted is in the form of an alpha particle (a helium ion) or subatomic particles called beta particles and neutrinos.

reductionism   The study of the component parts of a system for the sake of simplifying scientific analyses.

rhizosphere   An area of soil or other medium closely surrounding a plant root, where the plant secretes factors that support a robust microbial community.

ribosome   A complex of proteins and RNA molecules that synthesizes proteins in all living cells.

ribozyme   An RNA molecule that catalyzes chemical reactions, including the formation of other RNA molecules.

root nodule   A structure seen on the roots of leguminous plants specialized for symbiosis with nitrogen-fixing bacteria.

RNA   Ribonucleic acid. The nucleic acid seen as messenger RNA, transfer RNA, ribosomal RNA, and other types of functional RNA in the cells of all cellular organisms. Also the primary genetic information of some viruses containing RNA genomes instead of DNA genomes.

RNA polymerase   An enzyme in modern cells that rapidly produces RNA molecules from a DNA template, driving the process of transcription.

RNA replicase   Any enzyme that catalyzes the replication of RNA molecules.

S

scavenger   An organism that feeds on dead organisms.

SCOBY   A symbiotic colony of bacteria and yeast. The acronym is used as a name for the floating multispecies biofilm formed during the fermentation of kombucha.

selective sweep   A phenomenon in population genetics whereby a mutation or other genetic change that occurs in one individual sweeps across and becomes fixed within the entire population, eliminating the diversity at the genetic locus that might have existed before the sweep. A result of a beneficial mutation that increases the fitness of individuals carrying it.

selfish genetic element   Any entity containing genetic information that replicates within a host organism but does not contribute to the fitness or reproductive success of the host genome.

self-organization or self-assembly   The spontaneous emergence of order occurring whenever a less organized system becomes ordered without the influence of an external agent. A result of local interactions between subparts.

S-layer   A proteinaceous outer layer of the cell envelope seen in most archaea and in some bacteria.

sliding   Bacterial locomotion across a surface driven by colony expansion, extracellular-matrix production, and sometimes also a surfactant that reduces surface tension.

small subunit ribosomal RNA   A component of the ribosome highly conserved across all cells. Because small subunit ribosomal RNA evolves more slowly than other genes, its sequence can be compared among all extant species to determine how they are related to each other.

sociality   The degree to which individuals in a population associate and form cooperative groups.

sociobiology   The scientific study of social behaviors and organization among organisms and their societies.

sociomicrobiology   Sociobiology in the context of microbial species.

solar nebula   A cosmic cloud of dust and debris that coalesces to form solar systems.

solar system   Any system of planets and other objects bound by the gravity of one or several stars.

solar wind   A constant stream of charged particles produced by the Sun that moves across the Solar System at high speed.

spatial resolution   A measure of an optical system's ability to differentiate between discrete points or objects of a given size. The spatial resolution of the human eye determines what are the smallest objects that we can see. The spatial resolution of an optical imaging system such as a light microscope determines the level of detail within the magnified image.

spore   A small reproductive unit fundamental to the life cycle of many organisms including bacteria, fungi, and some plants. Bacterial spores are tough structures that allow an organism to survive extreme environmental conditions and reanimate at a later time under favorable conditions.

streamlined genome   The reduced genome an organism retains when previously useful genes are no longer needed. Based on the principle that there is an energy cost to copying DNA and producing molecules useless to the cell. Commonly seen in endosymbionts, ectosymbionts, and among interdependent microbial networks.

streptomycin   An antibiotic first isolated from the bacterium *Streptomyces griseus*. Streptomycin binds to bacterial ribosomes and interferes with protein synthesis (translation). Used clinically for the treatment of tuberculosis and some other infections caused by gram-negative bacteria.

stromatolite   A microbialite with a laminated or layered structure.

stylet   The piercing mouthpart of a tardigrade and other organisms.

supernova   The explosion of a star. Often followed by a shock wave as the star's matter moves through the cosmos at up to 30,000 kilometers per second.

superorganism   A level of biological organization emerging from many closely related individual organisms that itself has the properties of an organism. An example is a leaf-cutter ant colony, in which one reproductive female (the queen) lives alongside millions of sterile individuals divided into distinct functional groups that cannot survive prolonged periods on their own.

supervolcano   The most catastrophic class of volcanos, capable of ejecting over a trillion tons of material from Earth into the atmosphere. Caused when a trickle of magma from within the planet accumulates and builds pressure in massive pools beneath the crust. An example is the Yellowstone supervolcano, which last erupted about 630,000 years ago.

surfactant   Chemicals that reduce the surface tension between two liquids or between a liquid and a solid.

swarming   A collective behavior in which individual organisms or other entities aggregate and move together. With respect to microbes, used specifically to describe a type of group migration across a surface driven by flagella and aided by a surfactant.

symbiosis   When two or more species live together through an intimate and long-term biological interaction. Although typically thought of as mutualistic—benefiting both partner species—symbiosis can be parasitic. In such cases, one species is harmed for the benefit of the other. The condition is an obligate one if one or both partners depend on it and facultative when the species can live without aid.

symbiosome   A specialized area within plant or animal tissues where an endosymbiont is held. An evolved structure of the host organism that facilitates symbiosis.

syncytium   One contiguous cell with multiple nuclei.

synthesis   As a means of scientific analysis (in contrast to reductionism), the combining of knowledge gained from different and sometimes disparate disciplines in order to understand natural phenomena at a higher-order or systems level.

syntrophy   Also called cross-feeding, a common condition in which a number of organisms live together by consuming each other's metabolic products.

T

tail fiber   The leg-like fibers on some phages that aid them in attaching to host cells.

thermophilic   A heat-loving organism that lives at temperatures considered extremely hot relative to those preferred by other forms of life.

Titan   Saturn's largest moon. Confirmed to have stable reservoirs of liquid methane on its surface through observation with the Hubble Space Telescope and through fly-by missions. Predicted to be rich in the prebiotic chemicals that could have led to the emergence of life.

transcription   The process of copying genetic information encoded by DNA into RNA molecules, including messenger RNAs that are then translated into proteins.

transduction   Horizontal gene transfer as mediated by viruses.

transformation   Horizontal gene transfer as mediated by extracellular DNA. Transformation may occur naturally through an active cellular process or may be induced artificially in the laboratory.

translation   The process of decoding genetic information in RNA molecules (messenger RNA) into protein molecules that carry out structural or enzymatic functions.

tree of life   A metaphor for the evolution and diversity of life on Earth. The branches of the tree represent groups of organisms that pass down genetic information from parent to offspring. The metaphor was helpful in early thinking on evolution and remains as a term of convenience. However, it falls short of depicting biology on Earth as we know it today, because it accounts poorly for symbiosis and horizontal gene transfer.

tuberculosis   The human disease caused by the bacterium *Mycobacterium tuberculosis*, affecting primarily the lungs and sometimes also other tissues.

turbidity   Visible cloudiness in a liquid caused by a high density of particulate matter, such as many millions of individually microscopic microbial cells.

U

ultra-small cell   In reference to bacteria and archaea, any species with unusually small cells, often straddling or below the diffraction limit of light (about 100–300 nanometers). Also called ultramicrobacteria or ultramicroarchaea.

ultra-structure   A description of any structure within a cell. Also known as subcellular.

ultraviolet (radiation and light)   The part of the electromagnetic spectrum with wavelengths shorter than visible light and longer than X-rays, corresponding to wavelengths of 10–400 nanometers.

unicellular organism   A sometimes misleading term for organisms that are not multicellular or appear not to be multicellular, because the cells are physically separated from one another. The term is sometimes supported by the true biological context of the species in question and may accurately describe the minimal unit needed for reproduction. However, it should be avoided as a blanket term for bacteria and archaea because it does not account for pervasive unseen interactions among physically disconnected cells of the same species, including horizontal gene transfer, chemical communication, and other forms of communication.

universe   All matter and space. Everything that humans can perceive in nature and all that we will ever perceive in nature.

V

vacuole   A vesicle-like organelle within a cell that can have many functions. An example is the food vacuole. In paramecia and other bacterivorous protists, food vacuoles contain digestive enzymes that break down prey microbes into nutrients.

valve   In microbiology, the individual halves of diatom shells (also called frustules), made of silicon dioxide.

van Gogh bundle   A microscopic, multicellular structure observed during the migration behavior of *Bacillus subtilis* bacteria across an agar medium, which superficially resembles the swirling patterns seen in Vincent van Gogh's *The Starry Night*.

vertical inheritance   The passing down of genetic information from one generation to the next. Typically thought of as passing from parent to offspring.

visible light or visible spectrum   The part of the electromagnetic spectrum seen with the human eye, corresponding to wavelengths of 400–700 nanometers.

vesicle   A compartment defined by a lipid bilayer surrounding internal fluid and surrounded by external fluid. Can be intracellular or extracellular.

visual acuity   The spatial resolution of an animal's visual system. Visual acuity determines the limits to perceiving objects given their sizes and distances from the observer.

W

wavelength   The physical distance between the peaks of a wave, or the spatial period of that wave. Wavelength determines the fundamental properties of the various types of electromagnetic radiation.

web of life   A metaphor for thinking about the evolution and diversity of life on Earth, similar to and expanding upon the tree of life. As in the tree of life, the structure or branches of the comprise groups of organisms that pass down genetic information from parent to offspring. Unlike the tree of life, the web visually conveys endosymbiosis and horizontal gene transfer as fundamental evolutionary forces.

Z

zircon crystal   A crystal composed of a mineral with the chemical formula $ZrSiO_4$. Notably useful for radiocarbon dating.

zooplankton   Microscopic animals, larvae of larger animals, and other small heterotrophic species that drift in oceans, lakes, and other bodies of water, along with phytoplankton.

zooxantheallae   Photosynthetic algae that live within coral tissues as symbionts.

Domains of Life

Macroscopic and microscopic life on Earth are divided into groups of related organisms. Listed from the broadest to the least broad, these are: domain, kingdom, phylum, class, order, family, genus, and species. The term "strain" is often used to describe specific varieties within species. The species featured in the book are divided by the domains Bacteria, Archaea, and Eukarya and by the additional category of noncellular biological entities, the viruses.

Archaea A domain of life including mostly microscopic organisms and their macroscopic manifestations. Species of archaea were likely observed for as long as most bacteria have been, but they were not recognized as a distinct group until a landmark study by Carl Woese and George Fox published in 1977. Uncapitalized when not directly referring to the domain itself.

Bacteria A domain of life including mostly microscopic organisms and their macroscopic manifestations and some species with visible single cells. Includes the "candidate phyla radiation": a diverse group of species found in almost all environments but only recently discovered because they are so small. Uncapitalized when not referencing the domain.

Eukarya A domain of life including diverse microscopic species and their macroscopic manifestations, along with most large multicellular organisms that fill our everyday perception of the world, including plants and humans. Eukaryotes arose from the endosymbiosis of a bacterium within a host archaeal cell billions of years ago and have been shaped by additional endosymbiotic events since then, such as the cyanobacterium that became the chloroplast seen in algae and plants. Strictly speaking, Eukarya is a special group that, according to the most recent phylogenetic analyses, lies within the domain Archaea.

Viruses These always-microscopic biological entities infect cellular organisms and depend upon their hosts in order to replicate. According to some strict definitions of life, viruses are not alive, yet they are not completely nonliving, as even the simplest viruses have emergent properties that the chemical entities comprising them do not have. Viruses have likely existed for at least as long as cellular life. However, unlike Bacteria, Archaea, and Eukarya, viruses cannot be grouped and analyzed as a cohesive evolutionary domain of life. One reason is that virus genomes (composed of either DNA or RNA) are mosaic and in constant flux, recombining their genetic material with one another and re-assorting their own genome to the extent that, at longer time scales, their evolution is untraceable.

FURTHER READING

The following books and articles provide a deeper view into the literature concerning many of the species, topics, and scientists discussed in each chapter. Open-access online research articles were favored whenever possible. The list includes a mixture of historic references and recent advances, categorized by scientific content.

1. FROM AN ANCIENT CHALK GRAVEYARD

Antoni van Leeuwenhoek and the Discovery of Microbes

Dobell, C. 1932. *Antony van Leeuwenhoek and His "Little Animals."* New York: Harcourt, Brace and Company.

> The following quotation from Leeuwenhoek in 1674 details his studies of chalk: "About six years ago, being in England, out of curiosity, and seeing the great chalk cliffs and chalky lands at Gravesend and Rochester, it oft-times set me a-thinking; and at the same time I also tried to penetrate the parts of the chalk. At last I observed that chalk consisteth of very small transparent particles; and these transparent particles lying one upon another, is, methinks now, the reason why chalk is white."

Gest, H. 2004. The discovery of microorganisms by Robert Hooke and Antoni van Leeuwenhoek, Fellows of the Royal Society. *Notes and Records of the Royal Society* 58 (2):187–201.

Snyder, L. J. 2015. *Eye of the Beholder: Johannes Vermeer, Antoni van Leeuwenhoek, and the Reinvention of Seeing.* New York: W. W. Norton & Company.

Coccolithophores

Falkowski, P. G., M. E. Katz, A. H. Knoll, A. Quigg, J. A. Raven, O. Schofield, and F. J. R. Taylor. 2004. The evolution of modern eukaryotic phytoplankton. *Science* 305:354–60.

Hooke, Robert. 1665. *Micrographia.* Royal Society. Available at http://www.gutenberg.org/ebooks/15491.

> A heavily illustrated book of "minute bodies made by magnifying glasses, with observations and inquiries thereupon." Hooke's tract contains the earliest use of the word "cell" and is credited as the first scientific best-seller.

Monteiro, F. M., L. T. Bach, C. Brownlee, P. Bown, R. E. M. Rickaby, A. J. Poulton, T. Tyrrell, L. Beaufort, S. Dutkiewicz, S. Gibbs, M. A. Gutowska, R. Lee, U. Riebesell, J. Young, and A. Ridgwell. 2016. Why marine phytoplankton calcify. *Science Advances* 2 (7):E1501822.

Raitsos, D. E., S. J. Lavender, Y. Pradhan, T. Tyrell, P. C. Reid, and M. Edwards. 2006. Coccolithophore bloom size variation in response to the regional environment of the subartic North Atlantic. *Limnology and Oceanography* 51 (5):2122–30.

Segev, E., T. P. Wyche, K. H. Kim, J. Petersen, C. Ellebrandt, H. Vlamakis, N. Barteneva, J. N. Paulson, L. Chai, J. Clardy, and R. Kolter. 2016. Dynamic metabolic exchange governs a marine algal-bacterial interaction. *Elife* 5.
     This study shows how *Emiliania huxleyi* interacts with another microbe within the marine ecosystem, the bacterium *Roseobacter.*

Vardi, A., L. Haramaty, B. A. Van Mooy, H. F. Fredricks, S. A. Kimmance, A. Larsen, and K. D. Bidle. 2012. Host-virus dynamics and subcellular controls of cell fate in a natural coccolithophore population. *Proceedings of the National Academy of Sciences USA* 109 (47):19327–32.
     Coccolithophore blooms are composed of not only algal cells but also transparent exopolymer particles. This study shows that *Emiliania huxleyi* produces transparent exopolymer particles in response to, and likely as a defense against, infection by coccolithoviruses.

Witty, M. 2011. The White Cliffs of Dover are an example of natural carbon sequestration. *Ecologia* 1 (1):23–30.

Effects of Microbes in Our Lives and on the Planet at Large

Gilbert, J. A., and J. D. Neufeld. 2014. Life in a world without microbes. *PLoS Biology* 12 (12):E1002020.

Microbial Biodiversity and Microbial Numbers

Locey, K. J., and J. T. Lennon. 2016. Scaling laws predict global microbial diversity. *Proceedings of the National Academy of Sciences USA* 113 (21):5970–75.
     This study predicts that there as many as one trillion microbial species on Earth.

Mark Welch, J. L., B. J. Rossetti, C. W. Rieken, F. E. Dewhirst, and G. G. Borisy. 2016. Biogeography of a human oral microbiome at the micron scale. *Proceedings of the National Academy of Sciences USA* 113 (6):E791–800.

The CLASI-FISH microscopy method used in this study provides an exceptionally detailed view of microbial communities in human dental plaque—the same plaque first examined by Antoni van Leeuwenhoek in the late 1600s.

Microbiology by numbers. 2011. *Nature Reviews Microbiology* 9 (9):628.
A handful of interesting microbial numbers reported (all ballpark figures):

- There are 13×10^{28} bacteria in the oceans.

- There are 1×10^{31} viruses on Earth. Stretched end to end, these would be greater than 100 million light years in length.

- There are 1×10^{23} viral infections every second in the ocean, clearing 20–40 percent of all bacterial cells in the ocean every day.

- There are 1×10^{9} microbes in a teaspoon of soil, roughly the same as the number of humans living in Africa.

- There are 1×10^{11} bacteria in every gram of dental plaque, roughly the same as the number of people that have ever lived.

- There is 1 kilogram of bacteria in the human gut.

- There are 958,980 atoms in every Simian virus particle.

Ultra-Small Bacteria

Brown, C. T., L. A. Hug, B. C. Thomas, I. Sharon, C. J. Castelle, A. Singh, M. J. Wilkins, K. C. Wrighton, K. H. Williams, and J. F. Banfield. 2015. Unusual biology across a group comprising more than 15% of domain *Bacteria*. *Nature* 523 (7559):208–11.

Luef, B., K. R. Frischkorn, K. C. Wrighton, H-Y. N. Holman, G. Birarda, B. C. Thomas, A. Singh, K. H. Williams, C. E. Siegerist, S. G. Tringe, K. H. Downing, L. R. Comolli, and J. F. Banfield. 2015. Diverse uncultivated ultra-small bacterial cells in groundwater. *Nature Communications* 6:6372.

The Tree of Life and Endosymbiosis

Hug, L. A., B. J. Baker, K. Anantharaman, C. T. Brown, A. J. Probst, C. J. Castelle, C. N. Butterfield, A. W. Hernsdorf, Y. Amano, K. Ise, Y. Suzuki, N. Dudek, D. A. Relman, K. M. Finstad, R. Amundson, B. C. Thomas, and J. F. Banfield. 2016. A new view of the tree of life. *Nature Microbiology* 1:16048.

Margulis, L., and J. F. Stolz. 1984. Cell symbiosis theory: Status and implications for the fossil record. *Advances in Space Research* 4 (12):195–201.

Pace, N. R. 1997. A molecular view of microbial diversity and the biosphere. *Science* 276 (5313):734–40.

Spang, A., J. H. Saw, S. L. Jørgensen, K. Zaremba-Niedzwiedzka, J. Martijn, A. E. Lind, R. van Eijk, C. Schleper, L. Guy, and T. J. Ettema. 2015. Complex archaea that bridge the gap between prokaryotes and eukaryotes. *Nature* 521 (7551):173–79.

Williams, T. A., P. G. Foster, C. J. Cox, and T. M. Embley. 2013. An archaeal origin of eukaryotes supports only two primary domains of life. *Nature* 504 (7479):231–36.

Woese, C. R., and G. E. Fox. 1977. Phylogenetic structure of the prokaryotic domain: The primary kingdoms. *Proceedings of the National Academy of Sciences USA* 74 (11):5088–90.

Zaremba-Niedzwiedzka, K., E. F. Caceres, J. H. Saw, D. Bäckström, L. Juzokaite, E. Van-caester, K. W. Seitz, K. Anantharaman, P. Starnawski, K. U. Kjeldsen, M. B. Stott, T. Nunoura, J. F. Banfield, A. Schramm, B. J. Baker, A. Spang, and T. J. Ettema. 2017. Asgard archaea illuminate the origin of eukaryotic cellular complexity. *Nature* 541 (7637):353–58.

Bacterial and Archaeal Cell Biology

Albers, S. V., and B. H. Meyer. 2011. The archaeal cell envelope. *Nature Reviews Microbiology* 9 (6):414–26.
    A review of the archaeal cell wall and the archaeal flagellum analog, the archaellum.

Cossart, P., and A. Lebreton. 2014. A trip in the "New Microbiology" with the bacterial pathogen *Listeria monocytogenes*. *FEBS Letters* 588 (15):2437–45.

Moffitt, J. R., S. Pandey, A. N. Boettiger, S. Wang, and X. Zhuang. 2016. Spatial organization shapes the turnover of a bacterial transcriptome. *Elife* 5.
    This study explores one example of the complex biology of bacterial and archaeal cells, showing that the process of transcription is spatially organized within the bacterial cell.

2. To the Heartbeat of Earth

The Formation of Earth and the Habitability of the Early Planet

Abramov, O., and S. J. Mojzsis. 2009. Microbial habitability of the Hadean Earth during the late heavy bombardment. *Nature* 459:419–22.
    This study models asteroid impacts during the Hadean period and argues that these did not increase the temperature of Earth's lithosphere enough to sterilize life.

Bell, E. A., P. Boehnke, T. M. Harrison, and W. L. Mao. 2015. Potentially biogenic carbon preserved in a 4.1 billion-year-old zircon. *Proceedings of the National Academy of Sciences USA* 112 (47):14518–21.

Boehnkea, P., and T. M. Harrison. 2016. Illusory late heavy bombardments. *Proceedings of the National Academy of Sciences USA* 113 (39):10802–06.

Bowring, S. A., I. S. Williams, and W. Compston. 1989. 3.96 Ga gneisses from the Slave Province, Northwest Territories, Canada. *Geology* 17 (11):971–75.
  This study calls into question the late-heavy bombardment hypothesis.

Hazen, R. M. 2012. *The Story of Earth: The First 4.5 Billion Years, from Stardust to Living Planet.* New York: Viking.

KamLAND Collaboration. 2011. Partial radiogenic heat model for Earth revealed by geoneutrino measurements. *Nature Geoscience* 4:647–51.
  This paper finds that a significant portion of Earth's internal heat is most likely primordial heat, left over from Earth's formation.

Merrill, R. T. 2010. *Our Magnetic Earth: The Science of Geomagnetism.* Chicago: The University of Chicago Press.

Reimink, J. R., T. Chacko, R. A. Stern, and L. M. Heaman. 2013. Earth's earliest evolved crust generated in an Iceland-like setting. *Nature Geoscience* 7:529–33.

Valley, J. W., A. J. Cavosie, T. Ushikubo, D. A. Reinhard, D. F. Lawrence, D. J. Larson, P. H. Clifton, T. F. Kelly, S. A. Wilde, D. E. Moser, and M. J. Spicuzza. 2014. Hadean age for a post-magma-ocean zircon confirmed by atom-probe tomography. *Nature Geoscience* 7:219–23.

The Origin of Life

Blain, J. C., and J. W. Szostak. 2014. Progress toward synthetic cells. *Annual Review of Biochemistry* 83:615–40.

Horning, D. P., and G. F. Joyce. 2016. Amplification of RNA by an RNA polymerase ribozyme. *Proceedings of the National Academy of Sciences USA* 113 (35):9786–91.

Mulkidjanian, A. Y., A. Y. Bychkov, D. V. Dibrova, M. Y. Galperin, and E. V. Koonin. 2012. Origin of first cells at terrestrial, anoxic geothermal fields. *Proceedings of the National Academy of Sciences USA* 109 (14):E821–30.

Patel, B. H., C. Percivalle, D. J. Ritson, C. D. Duffy, and J. D. Sutherland. 2015. Common origins of RNA, protein and lipid precursors in a cyanosulfidic protometabolism. *Nature Chemistry* 7 (4):301–07.

Robertson, M. P., and G. F. Joyce. 2012. The origins of the RNA world. *Cold Spring Harbor Perspectives in Biology* 4 (5).

Szostak, J. W. 2016. On the origin of life. *Medicina* 76 (4):199–203.

Weiss, M. C., F. L. Sousa, N. Mrnjavac, S. Neukirchen, M. Roettger, S. Nelson-Sathi, and W. F. Martin. 2016. The physiology and habitat of the last universal common ancestor. *Nature Microbiology* 1 (9):16116.

Woese, C. 1967. *The Genetic Code: The Molecular Basis for Genetic Expression*. New York: Harper & Row.

Bacteria that Sense Magnetic Fields

Rahn-Lee, L., and A. Komeili. 2013. The magnetosome model: Insights into the mechanisms of bacterial biomineralization. *Frontiers in Microbiology* 4:352.

Microbial Signatures in the Geological Record

Bosak, T., A. H. Knoll, and A. P. Petroff. 2013. The meaning of stromatolites. *Annual Review of Earth and Planetary Sciences* 41:21–44.

Eisenberg, L. 2003. Giant stromatolites and a supersurface in the Navajo Sandstone, Capitol Reef National Park, Utah. *Geology* 31 (2):111–14.

Kappler, A., C. Pasquero, K. O. Konhauser, and D. K. Newman. 2005. Deposition of banded iron formations by anoxygenic phototrophic Fe(II)-oxidizing bacteria. *Geology* 33 (11):865–68.

Nutman, A. P., V. C. Bennett, C. R. L. Friend, M. J. Van Kranendonk, and A. R. Chivas. 2016. Rapid emergence of life shown by discovery of 3,700-million-year-old microbial structures. *Nature* 537 (7621):535–38.

Raup, O. B., R. L. Earhart, J. W. Whipple, P. E. Carrara. 1983. *Geology along Going-to-the-Sun-Road, Glacier National Park, Montana*. West Glacier, MT: Glacier National History Association.

Wacey, D. 2010. Stromatolites in the 3400 Ma Strelley Pool Formation, Western Australia: Examining biogenicity from the macro- to the nano-scale. *Astrobiology* 10 (4): 381–95.

Modern Microbial Communities in Yellowstone, Great Salt Lake, and Other Extreme Environments

Barns, S. M., R. E. Fundyga, M. W. Jeffries, and N. R. Pace. 1994. Remarkable archaeal diversity detected in a Yellowstone National Park hot spring environment. *Proceedings of the National Academy of Sciences USA* 91 (5):1609–13.

Berelson, W. M., F. A. Corsetti, C. Pepe-Ranney, D. E. Hammond, W. Beaumont, and J. R. Spear. 2011. Hot spring siliceous stromatolites from Yellowstone National Park: Assessing growth rate and laminae formation. *Geobiology* 9 (5):411–24.

Huang, H., F. Lin, B. Schmandt, J. Farrell, R. B. Smith, and V. C. Tsai. 2015. The Yellowstone magmatic system from the mantle plume to the upper crust. *Science* 348 (6236):773–76.

Klatt, C. G., J. M. Wood, D. B. Rusch, M. M. Bateson, N. Hamamura, J. F. Heidelberg, A. R. Grossman, D. Bhaya, F. M. Cohan, M. Kuhl, D. A. Bryant, and D. M. Ward. 2011. Community ecology of hot spring cyanobacterial mats: Predominant populations and their functional potential. *ISME Journal* 5 (8):1262–78.

Oren, A. 1994. The ecology of the extremely halophilic archaea. *FEMS Microbiology Reviews* 13 (4):415–40.

Pace, A., R. Bourillot, A. Bouton, E. Vennin, S. Galaup, I. Bundeleva, P. Patrier, C. Dupraz, C. Thomazo, P. Sansjofre, Y. Yokoyama, M. Franceschi, Y. Anguy, L. Pigot, A. Virgone, and P. T. Visscher. 2016. Microbial and diagenetic steps leading to the mineralisation of Great Salt Lake microbialites. *Scientific Reports* 6:31495.

Rothschild, L. J., and R. L. Mancinelli. 2001. Life in extreme environments. *Nature* 409 (6823):1092–101.

Diatoms and Their Glass Shells

Bradbury, J. 2004. Nature's nanotechnologists: Unveiling the secrets of diatoms. *PLoS Biology* 2 (10):e306.
> This feature article includes some key facts about diatoms. Diatoms appeared 180 million years ago and diversified into about 100,000 species of two major types: centric (with a glass shell that has circular symmetry) and pennate (with bilateral symmetry). They range in size from a few micrometers to one millimeter.

The Gut-Brain Axis

Andrey Smith, P. 2015. The tantalizing links between gut microbes and the brain. *Nature* 526 (7573):312–14.

3. Under Celia Thaxter's Garden

On the Lives of Celia and Roland Thaxter

Curray, D. P. 1990. *Childe Hassam: An Island Garden Revisited*. New York: W.W. Norton.

Mandel, N. H. 2004. *Beyond the Garden Gate: The Life of Celia Laighton Thaxter*. Lebanon, NH: University Press of New England

Thaxter, Celia. 1873. *Among the Isles of Shoals*. Boston: James R. Osgood and Co.

Thaxter, Celia. 1894. *An Island Garden*. Boston: Houghton, Mifflin & Co.

The Soil Microbial Ecosystem

Dance, A. 2008. Soil ecology: What lies beneath. *Nature* 455:724–25.

Gerber, N. N., and H. A. Lechevalier. 1965. Geosmin, an earthy-smelling substance isolated from actinomycetes. *Applied Microbiology* 13 (6):935–38.

Horton, T. R., and T. D. Bruns. 2001. The molecular revolution in ectomycorrhizal ecology: Peeking into the black-box. *Molecular Ecology* 10 (8):1855–71.

Roesch, L. F., R. R. Fulthorpe, A. Riva, G. Casella, A. K. Hadwin, A. D. Kent, S. H. Daroub, F. A. Camargo, W. G. Farmerie, and E. W. Triplett. 2007. Pyrosequencing enumerates and contrasts soil microbial diversity. *ISME Journal* 1 (4):283–90.

Schloss, P. D., and J. Handelsman. 2006. Toward a census of bacteria in soil. *PLoS Computational Biology* 2 (7):e92.

Simard, S. W., D. A. Perry, M. D. Jones, D. D. Myrold, D. M. Durall, and R. Molina. 1997. Net transfer of carbon between ectomycorrhizal tree species in the field. *Nature* 388 (7).

Vitousek, P. M., D. N. L. Menge, S. C. Reed, and C. C. Cleveland. 2013. Biological nitrogen fixation: Rates, patterns and ecological controls in terrestrial ecosystems. *Philosophical Transactions of the Royal Society of London Series B* 368 (1621):20130119.

The "Humongous Fungus" in Oregon

Ferguson, B. A., T. A. Dreisbach, C. G. Parks, G. M. Filip, and C. L. Schmitt. 2003. Coarse-scale population structure of pathogenic *Armillaria* species in a mixed-conifer forest in the Blue Mountains of northeast Oregon. *Canadian Journal of Forest Research* 33 (4):612–23.

Gould, S. J. 1992. A humongous fungus among us. *Natural History* 101 (7):10.

Salt Marsh Microbial Mats

Armitage, D. W., K. L. Gallagher, N. D. Youngblut, D. H. Buckley, and S. H. Zinder. 2012. Millimeter-scale patterns of phylogenetic and trait diversity in a salt marsh microbial mat. *Frontiers in Microbiology* 3:293.

Nicholson, J. A., J. F. Stolz, and B. K. Pierson. 1987. Structure of a microbial mat at Great Sippewissett Marsh, Cape Cod, Massachusetts. *FEMS Microbiology Letters* 45 (6):343–64.

Wilbanks, E. G., U. Jaekel, V. Salman, P. T. Humphrey, J. A. Eisen, M. T. Facciotti, D. H. Buckley, S. H. Zinder, G. K. Druschel, D. A. Fike, and V. J. Orphan. 2014. Microscale sulfur cycling in the phototrophic pink berry consortia of the Sippewissett Salt Marsh. *Environmental Microbiolgy* 16 (11):3398–415.

The Symbioses within Lichen

Grube, M., T. Cernava, J. Soh, S. Fuchs, I. Aschenbrenner, C. Lassek, U. Wegner, D. Becher, K. Riedel, C. W. Sensen, and G. Berg. 2015. Exploring functional contexts of symbiotic sustain within lichen-associated bacteria by comparative omics. *ISME Journal* 9 (2):412–24.

Scheidegger, C. 2016. As thick as three in a bed. *Molecular Ecology* 25 (14):3261–63.

Tardigrades, Rotifers, and Other Micro-Animals of the Moss Microbiome

Hashimoto, T., D. D. Horikawa, Y. Saito, H. Kuwahara, H. Kozuka-Hata, T. Shin-I, Y. Minakuchi, K. Ohishi, A. Motoyama, T. Aizu, A. Enomoto, K. Kondo, S. Tanaka, Y. Hara, S. Koshikawa, H. Sagara, T. Miura, S. Yokobori, K. Miyagawa, Y. Suzuki, T. Kubo, M. Oyama, Y. Kohara, A. Fujiyama, K. Arakawa, T. Katayama, A. Toyoda, and T. Kunieda. 2016. Extremotolerant tardigrade genome and improved radiotolerance of human cultured cells by tardigrade-unique protein. *Nature Communications* 7:12808.

Kostka, J. E., D. J. Weston, J. B. Glass, E. A. Lilleskov, A. J. Shaw, and M. R. Turetsky. 2016. The *Sphagnum* microbiome: New insights from an ancient plant lineage. *New Phytologist* 211 (1):57–64.

Vlčková, Š., J. Linhart, and V. Uvíra. 2002. Permanent and temporary meiofauna of an aquatic moss *Fontinalis antipyretica* Hedw. *Acta Universitatis Palackianae Olomucensis Facultas Rerum Naturalium Biologica* 39–40:131–40.

The Leaf-Surface Microbiome

Lambais, M. R., D. E. Crowley, J. C. Cury, R. C. Büll, and R. R. Rodrigues. 2006. Bacterial diversity in tree canopies of the Atlantic forest. *Science* 312 (5782):1917.

Microbial Diversity and Activities in the Ocean

Levitan, O., J. Dinamarca, G. Hochman, and P. G. Falkowski. 2014. Diatoms: A fossil fuel of the future. *Trends in Biotechnology* 32 (3):117–24.
  Diatoms in the ocean and other environments fix carbon efficiently and are considered strong candidates for the production of next-generation biofuels.

Moran, M. A. 2015. The global ocean microbiome. *Science* 350 (6266):aac8455.

Reisch, C. R., M. A. Moran, and W. B. Whitman. 2011. Bacterial catabolism of dimethylsulfoniopropionate (DMSP). *Frontiers of Microbiology* 2:172.

The Human Microbiome and the Microbiology of the Built Environment

Gaci, N., G. Borrel, W. Tottey, P. W. O'Toole, and J. F. Brugère. 2014. Archaea and the human gut: New beginning of an old story. *World Journal of Gastroenterology* 20 (43):16062–78

Grice, E. A., and J. A. Segre. 2011. The skin microbiome. *Nature Reviews Microbiology* 9 (4):244–53.

Kelly, S. T., and J. A. Gilbert. 2013. Studying the microbiology of the indoor environment. *Genome Biology* 14 (2):202.

Kembel, S. W., E. Jones, J. Kline, D. Northcutt, J. Stenson, A. M. Womack, B. J. Bohannan, G. Z. Brown, and J. L. Green. 2012. Architectural design influences the diversity and structure of the built environment microbiome. *ISME Journal* 6 (8):1469–79.

Stephen Jay Gould's "Planet of the Bacteria" and Comparison to E. O. Wilson's Ant Counts

Gould, S. J. 1996. Planet of the bacteria. *Washington Post,* November 13: H1.

Wilson, E. O. 1987. The arboreal ant fauna of Peruvian Amazon forests: A first assessment. *Biotropica* 19 (3):245–51.

4. Intelligent Slime

Physarum polycephalum and Semi-Intelligent Behavior

Howard, F. L. 1931. The life history of *Physarum polycephalum*. *American Journal of Botany* 18 (2):116–33.
  This was the first detailed study of *Physarum polycephalum*.

Reid, C. R., M. Beekman, T. Latty, and A. Dussutour. 2013. Amoeboid organism uses extracellular secretions to make smart foraging decisions. *Behavioral Ecology* 24 (4):812–18.

Reid, C. R., and T. Latty. 2016. Collective behavior and swarm intelligence in slime moulds. *FEMS Microbiology Reviews* 40 (6):798–806.
  This review includes a description of the biological units known as biochemical oscillators.

Tero, A., R. Kobayashi, and T. Nakagaki. 2006. *Physarum* solver: A biologically inspired method of road-network navigation. *Physica A: Statistical Mechanics and Its Applications* 363 (1):115–19.

Vogel, D., and A. Dussutour. 2016. Direct transfer of learned behavior via cell fusion in non-neural organisms. *Proceedings of the Royal Society B: Biological Sciences* 283 (1845).

Whiting, J. G., J. Jones, L. Bull, M. Levin, and A. Adamatzky. 2016. Towards a *Physarum* learning chip. *Scientific Reports* 6:19948.
  This paper describes one of many potential applications of the basic science of slime molds.

Biofilms

Chimileski, S., M. J. Franklin, and R. T. Papke. 2014. Biofilms formed by the archaeon *Haloferax volcanii* exhibit cellular differentiation and social motility, and facilitate horizontal gene transfer. *BMC Biology* 12 (65).
  This study describes biofilm formation in an archaeal species. Biofilms mostly have been studied in bacteria.

Dietrich, L. E. P., C. Okegbe, A. Price-Whelan, H. Sakhtah, R. C. Hunter, and D. K. Newman. 2013. Bacterial community morphogenesis is intimately linked to the intracellular redox state. *Journal of Bacteriology* 195 (7):1371–80.
  This study uncovers mechanisms underpinning the intricate wrinkled patterns in *Pseudomonas* colony biofilms.

Flemming, H. C., T. R. Neu, and D. J. Wozniak. 2007. The EPS matrix: The "house of biofilm cells." *Journal of Bacteriology* 189 (22):7945–47.

Henrici, A. T. 1933. Studies of freshwater bacteria. I. A direct microscopic technique. *Journal of Bacteriology* 25 (3):277–87.

Zobell, C. E. 1943. The effect of solid surfaces upon bacterial activity. *Journal of Bacteriology* 46 (1):39–56.

Ferdinand Cohn and Robert Koch—Early Investigators of *Bacillus* Biology

Blevins, S. M., and M. S. Bronze. 2010. Robert Koch and the "golden age" of bacteriology. *International Journal of Infectious Diseases* 14 (9):e744–51.

Drews, G. 2000. The roots of microbiology and the influence of Ferdinand Cohn on microbiology of the 19th century. *FEMS Microbiology Reviews* 24 (3):225–49.

Microbial Multicellularity

Aguilar, C., H. Vlamakis, R. Losick, and R. Kolter. 2007. Thinking about *Bacillus subtilis* as a multicellular organism. *Current Opinion in Microbiology* 10 (6):638–43.
    This paper includes citations and further information on Cohn and Koch's 1877 studies on *Bacillus.*

Kysela, D. T., A. M. Randich, P. D. Caccamo, and Y. V. Brun. 2016. Diversity takes shape: Understanding the mechanistic and adaptive basis of bacterial morphology. *PLoS Biology* 14 (10):e1002565.

Shapiro, J. A. 1988. Bacteria as multicellular organisms. *Scientific American* 258 (6): 82–89.

van Gestel, J., H. Vlamakis, and R. Kolter. 2015. From cell differentiation to cell collectives: *Bacillus subtilis* uses division of labor to migrate. *PLoS Biology* 13 (4):e1002141.

Vlamakis, H., C. Aguilar, R. Losick, and R. Kolter. 2008. Control of cell fate by the formation of an architecturally complex bacterial community. *Genes & Development* 22 (7):945–53.

Sociobiology as Applied to Microbiology (or "Sociomicrobiology")

Nadell, C. D., J. B. Xavier, and K. R. Foster. 2009. The sociobiology of biofilms. *FEMS Microbiology Reviews* 33 (1):206–24.

Oliveira, N. M., E. Martinez-Garcia, J. Xavier, W. M. Durham, R. Kolter, W. Kim, and K. R. Foster. 2015. Biofilm formation as a response to ecological competition. *PLoS Biology* 13 (7):e1002191.

Oliveira, N. M., R. Niehus, and K. R. Foster. 2014. Evolutionary limits to cooperation in microbial communities. *Proceedings of the National Academy of Sciences USA* 111 (50):17941–46.

Parsek, M. R., and E. P. Greenberg. 2005. Sociomicrobiology: The connections between quorum sensing and biofilms. *Trends in Microbiology* 13 (1):27–33.

Mechanisms of Microbial Cooperation and Competition

Alteri, C. J., S. D. Himpsl, S. R. Pickens, J. R. Lindner, J. S. Zora, J. E. Miller, P. D. Arno, S. W. Straight, and H. L. Mobley. 2013. Multicellular bacteria deploy the type VI secretion system to preemptively strike neighboring cells. *PLoS Pathogens* 9 (9):e1003608.

Basler, M., B. T. Ho, and J. J. Mekalanos. 2013. Tit-for-tat: Type VI secretion system counterattack during bacterial cell-cell interactions. *Cell* 152 (4):884–94.

Stefanic, P., B. Kraigher, N. A. Lyons, R. Kolter, and I. Mandic-Mulec. 2015. Kin discrimination between sympatric *Bacillus subtilis* isolates. *Proceedings of the National Academy of Sciences USA* 112 (45): 14042–47.

How Microbes Communicate

El-Naggar, M. Y., G. Wanger, K. M. Leung, T. D. Yuzvinsky, G. Southam, J. Yang, W. M. Lau, K. H. Nealson, and Y. A. Gorby. 2010. Electrical transport along bacterial nanowires from *Shewanella oneidensis* MR-1. *Proceedings of the National Academy of Sciences USA* 107 (42):18127–31.
     A study of the electrical intercellular filaments known as nanowires.

Fuqua, W. C., S. C. Winans, and E. P. Greenberg. 1994. Quorum sensing in bacteria: The LuxR-LuxI family of cell density-responsive transcriptional regulators. *Journal of Bacteriology* 176 (2):269–75.
     The term *quorum sensing* was coined in this paper.

Konovalova, A., and L. Søgaard-Andersen. 2011. Close encounters: Contact-dependent interactions in bacteria. *Molecular Microbiology* 81 (2):297–301.

López, D., H. Vlamakis, R. Losick, and R. Kolter. 2009. Paracrine signaling in a bacterium. *Genes & Development* 23 (14):1631–38.

Waters, C. M., and B. L. Bassler. 2005. Quorum sensing: Cell-to-cell communication in bacteria. *Annual Review of Cell and Developmental Biology* 21:319–46.

Prochlorococcus Ecology, Evolution, and Collective Dynamics

Biller, S. J., P. M. Berube, D. Lindell, and S. W. Chisholm. 2015. *Prochlorococcus*: The structure and function of collective diversity. *Nature Reviews Microbiology* 13 (1):13–27.

Biller, S. J., F. Schubotz, S. E. Roggensack, A. W. Thompson, R. E. Summons, and S. W. Chisholm. 2014. Bacterial vesicles in marine ecosystems. *Science* 343 (6167):183–86.

Huang, S., S. Zhang, N. Jiao, and F. Chen. 2015. Marine cyanophages demonstrate biogeographic patterns throughout the global ocean. *Applied and Environmental Microbiology* 81 (1):441–52.

Sullivan, M. B., M. L. Coleman, P. Weigele, F. Rohwer, and S. W. Chisholm. 2005. Three *Prochlorococcus* cyanophage genomes: Signature features and ecological interpretations. *PLoS Biology* 3 (5):e144.

Waterbury, J. 2004. Little things matter a lot. *Oceanus* 43 (2): 12–16.
    This magazine article covers the discovery of *Prochlorococcus* and marine cyanobacteria's major contribution to global oxygen levels.

The Black Queen Hypothesis and the Interdependence of *Prochlorococcus* and *Pelagibacter*

Morris, J. J., R. E. Lenski, and E. R. Zinser. 2012. The Black Queen Hypothesis: Evolution of dependencies through adaptive gene loss. *mBio* 3 (2):e00036–12.

Morris, J. J., S. E. Papoulis, and R. E. Lenski. 2014. Coexistence of evolving bacteria stabilized by a shared Black Queen function. *Evolution* 68 (10):2960–71.

Myxobacteria and Their Advanced Social Behaviors

Berleman, J. E., T. Chumley, P. Cheung, and J. R. Kirby. 2006. Rippling is a predatory behavior in *Myxococcus xanthus*. *Journal of Bacteriology* 188 (16):5888–95.

Kaiser, D. 1993. Roland Thaxter's legacy and the origins of multicellular development. *Genetics* 135 (2):249–54.

Konovalova, A., T. Petters, and L. Søgaard-Andersen. 2010. Extracellular biology of *Myxococcus xanthus*. *FEMS Microbiology Reviews* 34 (2):89–106.

Thaxter, R. 1892. On the Myxobacteriaceæ, a new order of Schizomycetes. *Botanical Gazette* 17 (12):389–406.

Vassallo, C., D. T. Pathak, P. Cao, D. M. Zuckerman, E. Hoiczyk, and D. Wall. 2015. Cell rejuvenation and social behaviors promoted by LPS exchange in myxobacteria. *Proceedings of the National Academy of Sciences USA* 112 (22):E2939–46.

Collective Behavior

Gordon, D. M. 2014. The ecology of collective behavior. *PLoS Biology* 12 (3):e1001805.

Haeckel, E. *Art Forms in Nature.* 1974 [1904]. New York: Dover Publications.
    One hundred prints of microscopic and macroscopic organisms published between 1899 and 1904 under the German title *Kunstformen der Natur.* Many of the species portrayed were undescribed at the time. The book was influential for its contributions to biology, early twentieth-century design, architecture, and art. It remains popular for bridging the worlds of art and science.

5. TALES OF SYMBIOSIS

Llewelyn Williams's 1929 Trip to Peru as Part of the Captain Marshall Field Expeditions

Williams, L. 1936. *Woods of Northeastern Peru.* Chicago: Field Museum Press.

Williams, L. 1930. Medicine: Jungle surgery. *Time Magazine* 25 (17).
    The following passage recounts Williams's story of ant surgery. "When Indian warriors return home after inter-tribal clashes, tribeswomen anesthetize their wounds with ginger. Beside the 'doctor' is a jar containing a species of ferocious, strong-jawed ants. After drawing the lips of the wound together, the 'doctor' holds an ant close to the wound, lets it bite. If and when the mandibles strike on each side of the wound, the ant's body is snipped off. The death grip of the head holds the wound together."

Leaf-Cutter Ant Biology

Belt, T. 1874. *The Naturalist in Nicaragua.* London: John Murray.

Hölldobler, B., and E. O. Wilson. 2011. *The Leafcutter Ants: Civilization by Instincts.* New York: W.W. Norton.

Schultz, T. R., and S. G. Brady. 2008. Major evolutionary transitions in ant agriculture. *Proceedings of the National Academy of Sciences USA* 105 (14):5435–440.

The Symbiotic Network within a Leaf-Cutter Ant Fungus Garden

Currie, C. R., U. G. Mueller, and D. Malloch. 1999. The agricultural pathology of ant fungus gardens. *Proceedings of the National Academy of Sciences USA* 96 (14):7998–8002.

Currie, C. R., M. Poulsen, J. Mendenhall, J. J. Boomsma, and J. Billen. 2006. Coevolved crypts and exocrine glands support mutualistic bacteria in fungus-growing ants. *Science* 311 (5757):81–83.

Currie, C. R., J. A. Scott, R. C. Summerbell, and D. Malloch. 1999. Fungus-growing ants use antibiotic-producing bacteria to control garden parasites. *Nature* 398:701–04.

Gerardo, N. M., U. G. Mueller, S. L. Price, and C. R. Currie. 2004. Exploiting a mutualism: Parasite specialization on cultivars within the fungus-growing ant symbiosis. *Proceedings of the Royal Society B: Biological Sciences* 271 (1550):1791–98.

Haeder, S., R. Wirth, H. Herz, and D. Spiteller. 2009. Candicidin-producing *Streptomyces* support leaf-cutting ants to protect their fungus garden against the pathogenic fungus *Escovopsis*. *Proceedings of the National Academy of Sciences USA* 106 (12):4742–46.

Nygaard, S., H. Hu, C. Li, M. Schiøtt, Z. Chen, Z. Yang, Q. Xie, C. Ma, Y. Deng, R. B. Dikow, C. Rabeling, D. R. Nash, W. T. Wcislo, S. G. Brady, T. R. Schultz, G. Zhang, and J. J. Boomsma. 2016. Reciprocal genomic evolution in the ant-fungus agricultural symbiosis. *Nature Communications* 7:12233.

Pinto-Tomás, A. A., M. A. Anderson, G. Suen, D. M. Stevenson, F. S. T. Chu, W. W. Cleland, P. J. Weimer, and C. R. Currie. 2009. Symbiotic nitrogen fixation in the fungus gardens of leaf-cutter ants. *Science* 326 (5956):1120–23.

Endosymbiosis of Symbiodinium Dinoflagellates within Corals

Barott, K. L., A. A. Venn, S. O. Perez, S. Tambutté, and M. Tresguerres. 2015. Coral host cells acidify symbiotic algal microenvironment to promote photosynthesis. *Proceedings of the National Academy of Sciences USA* 112 (2):607–12.

Morden, C. W., and A. R. Sherwood. 2002. Continued evolutionary surprises among dinoflagellates. *Proceedings of the National Academy of Sciences USA* 99 (18):11558–60. A figure in this paper nicely illustrates the concepts of primary, secondary, and tertiary endosymbiosis.

Pochon, X., J. I. Montoya-Burgos, B. Stadelmann, and J. Pawlowski. 2006. Molecular phylogeny, evolutionary rates, and divergence timing of the symbiotic dinoflagellate genus *Symbiodinium*. *Molecular Phylogenetics and Evolution* 38 (1):20–30.

The Thin Line between Commensal and Pathogenic Symbionts of the Human Microbiome

Bomar, L., S. D. Brugger, B. H. Yost, S. S. Davies, and K. P. Lemon. 2016. *Corynebacterium accolens* releases antipneumococcal free fatty acids from human nostril and skin surface triacylglycerols. *mBio* 7 (1):e01725–15.

Brugger, S. D., L. Bomar, and K. P. Lemon. 2016. Commensal-pathogen interactions along the human nasal passages. *PLoS Pathogens* 12 (7):e1005633.

The Concept and Controversy of Holobionts and Hologenomes as Multispecies Biological Units

Bordenstein, S. R., and K. R. Theis. 2015. Host biology in light of the microbiome: Ten principles of holobionts and hologenomes. *PLoS Biology* 13 (8):e1002226.

Margulis, L., and R. Fester. 1991. *Symbiosis as a Source of Evolutionary Innovation: Speciation and Morphogenesis*. Cambridge, MA: MIT Press.

Moran, N. A., and D. B. Sloan. 2015. The hologenome concept: Helpful or hollow? *PLoS Biology* 13 (12):e1002311.

Shropshire, J. D., and S. R. Bordenstein. 2016. Speciation by symbiosis: The microbiome and behavior. *mBio* 7 (2):e01785-15.

The Ecological Role of Antibiotics

Clardy, J., M. A. Fischbach, and C. R. Currie. 2009. The natural history of antibiotics. *Current Biology* 19 (11):R437–41.

Romero, D., M. F. Traxler, D. López, and R. Kolter. 2011. Antibiotics as signal molecules. *Chemical Reviews* 111 (9):5492–505.

Antibiotic Mode of Action and the Development of Antibiotic Resistance

Davies, J., and D. Davies. 2010. Origins and evolution of antibiotic resistance. *Microbiology and Molecular Biology Reviews* 74 (3):417–33.

Davies, J., W. Gilbert, and L. Gorini. 1964. Streptomycin, suppression, and the code. *Proceedings of the National Academy of Sciences USA* 51 (5):883–90.

Demirci, H., F. Murphy, E. Murphy, S. T. Gregory, A. E. Dahlberg, and G. Jogl. 2013. A structural basis for streptomycin-induced misreading of the genetic code. *Nature Communications* 4:1355.
    This paper expands upon earlier work by using modern structural-biology methods to reveal the precise mechanism of action for streptomycin.

Flashlight Fish and Other Marine Life Harboring Bioluminescent Bacteria

Davis, M. P., J. S. Sparks, and W. L. Smith. 2016. Repeated and widespread evolution of bioluminescence in marine fishes. *PLoS One* 11 (6):e0155154.

Fenolio, D. 2016. *Life in the Dark: Illuminating Biodiversity in the Shadowy Haunts of Planet Earth*. Baltimore: Johns Hopkins University Press.

Hellinger, J., P. Jägers, M. Donner, F. Sutt, M. D. Mark, B. Senen, R. Tollrian, and S. Herlitze. 2017. The flashlight fish *Anomalops katoptron* uses bioluminescent light to detect prey in the dark. *PLoS One* 12 (2):e0170489.

Mammalian Evolution: What Were Human Ancestors Like When Ants Began Farming?

Halliday, T. J., P. Upchurch, and A. Goswami. 2017. Resolving the relationships of Paleocene placental mammals. *Biological Reviews of the Cambridge Philosophical Society* 92 (1):521–50.
> This review breaks down recent work in the field of mammalian evolution, including an emphasis on the small, insect-eating mammal thought to be the common ancestor of all placental mammals.

Wilford, J. N. 2013. Rat-size ancestor said to link man and beast. *New York Times* Feb. 8: A1.

Hawaiian Bobtail-Squid Symbiosis with *Vibrio* Bacteria

Koropatnick, T. A., J. T. Engle, M. A. Apicella, E. V. Stabb, W. E. Goldman, and M. J. McFall-Ngai. 2004. Microbial factor-mediated development in a host-bacterial mutualism. *Science* 306 (5699):1186–88.

McFall-Ngai, M. J., E. A. Heath-Heckman, A. A. Gillette, S. M. Peyer, and E. A. Harvie. 2012. The secret languages of coevolved symbioses: Insights from the *Euprymna scolopes–Vibrio fischeri* symbiosis. *Seminars in Immunology* 24 (1):3–8.

Nyholm, S. V., E. V. Stabb, E. G. Ruby, and M. J. McFall-Ngai. 2000. Establishment of an animal-bacterial association: Recruiting symbiotic *Vibrios* from the environment. *Proceedings of the National Academy of Sciences USA* 97 (18):10231–35.

The Discovery of the First Antibiotics and the Schatz-Waksman Controversy over Streptomycin

Alexander Fleming Laboratory Museum. 1999. The discovery and development of penicillin, 1928–1945. American Chemical Society and the Royal Society of Chemistry.

This electronic booklet describes how Alexander Fleming accidentally discovered penicillin. The antibiotic was produced by a fungal species, the mold *Penicillium notatum*, which contaminated one of Fleming's routine experiments while he was working at St. Mary's Hospital in London.

Meyers, M. A. 2012. *Prize Fight: The Race and the Rivalry to Be the First in Science.* London: Palgrave Macmillan.

Pringle, P. 2012. *Experiment Eleven: Dark Secrets behind the Discovery of a Wonder Drug.* New York: Walker Publishing Company.

Schatz, A. 1993. The True Story of the Discovery of Streptomycin. *Actinomycetes* 4 (2):27–39.

Comparing Ants and Humans as Superorganisms

Farji-Brener, A. G., and V. Werenkraut. 2015. A meta-analysis of leaf-cutting ant nest effects on soil fertility and plant performance. *Ecological Entomology* 40 (2):150–58.

Hölldobler, B., and E. O. Wilson. 2008. *The Superorganism: The Beauty, Elegance, and Strangeness of Insect Societies.* New York: W.W. Norton.

Wilson, E. O. 2012. *The Social Conquest of Earth.* New York: W.W. Norton.

6. On the Kitchen Counter

Cheese Microbial Communities

Boynton, P. J., and D. Greig. 2014. The ecology and evolution of non-domesticated *Saccharomyces* species. *Yeast* 31 (12):449–62.

Brückner, A., and M. Heethoff. 2016. Scent of a mite: Origin and chemical characterization of the lemon-like flavor of mite-ripened cheeses. *Experimental and Applied Acarology* 69 (3):249–61.

Bull, M. J., K. A. Jolley, J. E. Bray, M. Aerts, P. Vandamme, M. C. Maiden, J. R. Marchesi, and E. Mahenthiralingam. 2014. The domestication of the probiotic bacterium *Lactobacillus acidophilus*. *Scientific Reports* 4:7202.

Button, J. E., and R. J. Dutton. 2012. Cheese microbes. *Current Biology* 22 (15):R587–9.

Wolfe, B. E., J. E. Button, M. Santarelli, and R. J. Dutton. 2014. Cheese rind communities provide tractable systems for in situ and in vitro studies of microbial diversity. *Cell* 158 (2):422–33.

Domestication of Cheese Microbes

Cheeseman, K., J. Ropars, P. Renault, J. Dupont, J. Gouzy, A. Branca, A. L. Abraham, M. Ceppi, E. Conseiller, R. Debuchy, F. Malagnac, A. Goarin, P. Silar, S. Lacoste, E. Sallet, A. Bensimon, T. Giraud, and Y. Brygoo. 2014. Multiple recent horizontal transfers of a large genomic region in cheese making fungi. *Nature Communications* 5:2876.

Gillot, G., J. L. Jany, M. Coton, G. Le Floch, S. Debaets, J. Ropars, M. López-Villavicencio, J. Dupont, A. Branca, T. Giraud, and E. Coton. 2015. Insights into *Penicillium roqueforti* morphological and genetic diversity. *PLoS One* 10 (6):e0129849.

Roca, M. G., N. D. Read, and A. E. Wheals. 2005. Conidial anastomosis tubes in filamentous fungi. *FEMS Microbiology Letters* 249 (2):191–98.

Ropars, J., R. C. Rodríguez de la Vega, M. López-Villavicencio, J. Gouzy, E. Sallet, É. Dumas, S. Lacoste, R. Debuchy, J. Dupont, A. Branca, and T. Giraud. 2015. Adaptive horizontal gene transfers between multiple cheese-associated fungi. *Current Biology* 25 (19):2562–69.

Zimmer, C. 2015. That stinky cheese is a result of evolutionary overdrive. *New York Times* Sept. 29: D4.

Evidence of Ancient Fermentation in Human Food and Drink

Crown, P. L., and W. J. Hurst. 2009. Evidence of cacao use in the Prehispanic American Southwest. *Proceedings of the National Academy of Sciences USA* 106 (7):2110–13.

Hurst, W. J., S. M. Tarka, T. G. Powis, F. Valdez, and T. R. Hester. 2002. Archaeology: Cacao usage by the earliest Maya civilization. *Nature* 418:289–90.

McGovern, P. E., J. Zhang, J. Tang, Z. Zhang, G. R. Hall, R. A. Moreau, A. Nuñez, E. D. Butrym, M. P. Richards, C. Wang, G. Cheng, Z. Zhao, and C. Wang. 2004. Fermented beverages of pre- and proto-historic China. *Proceedings of the National Academy of Sciences USA* 101 (51):17593–98.

Powis, T. G., W. J. Hurst, M. Carmen Rodríguez, P. Ortíz C., M. Blake, D. Cheetham, M. D. Coe, and J. G. Hodgson. 2007. Oldest chocolate in the New World. *Antiquity* 81 (314).

Domestication and Genomic Evolution of Yeasts (*Saccharomyces cerevisiae*)

Fay, J. C., and J. A. Benavides. 2005. Evidence for domesticated and wild populations of *Saccharomyces cerevisiae. PLoS Genetics* 1 (1):e5.

Gallone, B., J. Steensels, T. Prahl, L. Soriaga, V. Saels, B. Herrera-Malaver, A. Merlevede, M. Roncoroni, K. Voordeckers, L. Miraglia, C. Teiling, B. Steffy, M. Taylor, A. Schwartz, T. Richardson, C. White, G. Baele, S. Maere, and K. J. Verstrepen. 2016. Domestication and divergence of *Saccharomyces cerevisiae* beer yeasts. *Cell* 166 (6):1397–410e16.

Hyma, K. E., S. M. Saerens, K. J. Verstrepen, and J. C. Fay. 2011. Divergence in wine characteristics produced by wild and domesticated strains of *Saccharomyces cerevisiae*. *FEMS Yeast Research* 11 (7):540–51.

Sicard, D., and J. L. Legras. 2011. Bread, beer and wine: Yeast domestication in the *Saccharomyces* sensu stricto complex. *Comptes Rendus Biologies* 334 (3):229–36.

Studying Microbial Evolution

Fox, J. W., and R. E. Lenski. 2015. From here to eternity—the theory and practice of a really long experiment. *PLoS Biology* 13 (6):e1002185.

Domestication and Genomic Evolution of *Aspergillus*

Geiser, D. M, J. I. Pitt, and J. W. Taylor. 1998. Cryptic speciation and recombination in the aflatoxin-producing fungus *Aspergillus flavus*. *Proceedings of the National Academy of Sciences USA* 95 (1):388–93.

Machida, M., O. Yamada, and K. Gomi. 2008. Genomics of *Aspergillus oryzae*: Learning from the history of Koji mold and exploration of its future. *DNA Research* 15 (4):173–83.

Domestication of *Agaricus bisporus* and Other Mushrooms

Genders, R. 1969. *Mushroom Growing for Everyone*. London: Faber.

Schaechter, E. 1997. *In the Company of Mushrooms: A Biologist's Tale*. Cambridge, MA: Harvard University Press.

Spencer, D. M. 1985. The mushroom—its history and importance. In *The Biology and Technology of the Cultivated Mushroom*, edited by P. B. Flegg, D. M. Spencer, and D. A. Woods, 1–8. New York: John Wiley and Sons.

Fermented Food Microbiology

Marsh, A. J., O. O'Sullivan, C. Hill, R. P. Ross, and P. D. Cotter. 2014. Sequence-based analysis of the bacterial and fungal compositions of multiple kombucha (tea fungus) samples. *Food Microbiology* 38:171–78.

This metagenomic analysis reveals the most common bacteria, yeast, and other microbial species found in kombucha biofilms.

Wolfe, B. E., and R. J. Dutton. 2015. Fermented foods as experimentally tractable microbial ecosystems. *Cell* 161 (1):49–55.
This review covers fermented fruit products (wine, chocolate, coffee), dairy products (yogurt, cheese, kefir), grains (beer, sake, soy sauce, miso, sourdough), meats (salami), and plants (kimchi, sauerkraut, kombucha).

Cacao Seed Fermentation during Chocolate Making

Meersman, E., J. Steensels, M. Mathawan, P. J. Wittocx, V. Saels, N. Struyf, H. Bernaert, G. Vrancken, and K. J. Verstrepen. 2013. Detailed analysis of the microbial population in Malaysian spontaneous cocoa pulp fermentations reveals a core and variable microbiota. *PLoS One* 8 (12):e81559.

Schwan, R. F., and A. E. Wheals. 2004. The microbiology of cocoa fermentation and its role in chocolate quality. *Critical Reviews in Food Science and Nutrition* 44 (4):205–21.

Horizontal Gene Transfer and Its Impacts on the Tree or Web of Life

Niehus, R., S. Mitri, A. G. Fletcher, and K. R. Foster. 2015. Migration and horizontal gene transfer divide microbial genomes into multiple niches. *Nature Communications* 6:8924.

Polz, M. F., E. J. Alm, and W. P. Hanage. 2013. Horizontal gene transfer and the evolution of bacterial and archaeal population structure. *Trends in Genetics* 29 (3):170–75.

Zimmer, C. 2008. Festooning the tree of life. *Discover.* http://blogs.discovermagazine.com/loom/2008/07/17/festooning-the-tree-of-life.
This article details the effort to superimpose horizontal gene-transfer between separate lineages within the tree of life on top of a phylogenetic diagram of life: one way to visualize the concept of a web of life.

7. There Is Life at the Edge of Sight

Edwin Hubble and the Discovery of Other Galaxies

Hubble, E. 1936. *Realm of the Nebulae.* New Haven: Yale University Press.
This was Hubble's major work based on observations from the Hooker Telescope. Even after discovering that Andromeda and other so-called nebulae were "island universes," Hubble preferred "nebulae" over the new term "galaxy," coined by his

academic rival Harlow Shapley. Hubble wrote in *Realm of the Nebulae*, "The term nebulae offers the values of tradition; the term galaxies, the glamour of romance."

A new edition of Hubble's book, published in 2013, includes forewords from physicists Sean Carroll and Robert Kirshner.

National Aeronautics and Space Administration. 2011. *Hubble 21: Science Year in Review*, edited by K. Hartnett. NASA Pub. 2011–12–271-GSFC. NASA Goddard Space Flight Center, Greenbelt, MD.

This electronic book from NASA contains a history of the Hubble Space Telescope's development (propelled by Lyman Spitzer and others) and discusses basic aspects of the telescope's design and capabilities. The chapter "Revisiting V1" details Hubble's measurements of the Cepheid variable star known as V1 and includes images from 2010–2011 to commemorate the famous observations.

Finding Exoplanets and Assessing Their Habitability

Batalha, N. M. 2014. Exploring exoplanet populations with NASA's Kepler Mission. *Proceedings of the National Academy of Sciences USA* 111 (35):12647–54.

Fressin, F., G. Torres, D. Charbonneau, S. T. Bryson, J. Christiansen, C. D. Dressing, J. M. Jenkins, L. M. Walkowicz, and N. M. Batalha. 2013. The false positive rate of *Kepler* and the occurrence of planets. *Astrophysical Journal* 766 (2):81.

Jenkins, J. M., J. D. Twicken, N. M. Batalha, D. A. Caldwell, W. D. Cochran, M. Endl, D. W. Latham, G. A. Esquerdo, S. Seader, A. Bieryla, E. Petigura, D. R. Ciardi, G. W. Marcy, H. Isaacson, D. Huber, J. F. Rowe, G. Torres, S. T. Bryson, L. Buchhave, I. Ramirez, A. Wolfgang, J. Li, J. R. Campbell, P. Tenenbaum, D. Sanderfer, C. E. Henze, J. H. Catanzarite, R. L. Gilliland, and W. J. Borucki. 2015. Discovery and validation of Kepler-452b: A 1.6 R\\TNT2295\\ Super Earth exoplanet in the habitable zone of a G2 star. *Astrophysical Journal* 150 (2):56.

Sasselov, D. 2012. *The Life of Super-Earths*. New York: Basic Books.

Sloan Digital Sky Survey. 2017. The elements of life mapped across the Milky Way by SDSS / APOGEE. http://www.sdss.org/press-releases/the-elements-of-life-mapped -across-the-milky-way-by-sdssapogee.

The Hubble Space Telescope Extreme Deep Field Image

Illingworth, G. D., D. Magee, P. A. Oesch, R. J. Bouwens, I. Labbé, M. Stiavelli, P. G. van Dokkum, M. Franx, M. Trenti, C. M. Carollo, and V. Gonzalez. 2013. The HST extreme

Deep Field (XDF): Combining all Acs and Wfc3/Ir Data on the Hudf region into the deepest field ever. *Astrophysical Journal Supplement Series* 209 (1):6.

Integrated Views of the Microbial World and Life on Earth

Bosch, T. C., and M. J. McFall-Ngai. 2011. Metaorganisms as the new frontier. *Zoology* (Jena, Germany) 114 (4):185–90.

Davies, J. 2009. Everything depends on everything else. *Clinical Microbiology and Infection* 15 Suppl. 1:1–4.
   The title of this paper refers to the credo of the Haida people of the Queen Charlotte Islands, coastal British Columbia. The lifestyle of the Haida is guided by the principle of interdependence between humans, animals, and the environment.

Doolittle, W. Ford. 2012. Microbial neopleomorphism. *Biology & Philosophy* 28 (2):351–78.
   This paper argues that our ideas about microbes have come "full circle." First, we realized that there was a microbial world, but we did not have the technology to understand anything about the biology of microbes. We thought that microbes were spontaneously formed, came from some kind of "primoridal ooze," and could rapidly change form (pleomorphism). Later, scientists spent decades trying to organize microbial species using much the same approach they had applied to animal species. This assumed stable species with consistent, well-defined properties. Now, thanks to technologies such as DNA sequencing, we have found that microbes in fact *can* rapidly change through horizontal gene transfer; that they are not always clearly divisible into distinct species, as they live in interdependent networks; and that, within these networks, species are somewhat interchangeable in terms of the functions they carry out.

Ereshefsky, M., and M. Pedroso. 2015. Rethinking evolutionary individuality. *Proceedings of the National Academy of Sciences USA* 112 (33):10126–32.

Gordon, J., M. Youle, N. Knowlton, F. Rohwer, and D. A. Relman. 2013. Superorganisms and holobionts. *Microbe* 8 (4):152–53.

McFall-Ngai, M., M. G. Hadfield, T. C. Bosch, H. V. Carey, T. Domazet-Lošo, A. E. Douglas, N. Dubilier, G. Eberl, T. Fukami, S. F. Gilbert, U. Hentschel, N. King, S. Kjelleberg, A. H. Knoll, N. Kremer, S. K. Mazmanian, J. L. Metcalf, K. Nealson, N. E. Pierce, J. F. Rawls, A. Reid, E. G. Ruby, M. Rumpho, J. G. Sanders, D. Tautz, and J. J. Wernegreen. 2013. Animals in a bacterial world, a new imperative for the life sciences. *Proceedings of the National Academy of Sciences USA* 110 (9):3229–36.

Zhu, C., T. O. Delmont, T. M. Vogel, and Y. Bromberg. 2015. Functional basis of microorganism classification. *PLoS Computational Biology* 11 (8):e1004472.

The Archaeal Ectosymbiont *Nanopusillus* from Yellowstone National Park

Wurch, L., R. J. Giannone, B. S. Belisle, C. Swift, S. Utturkar, R. L. Hettich, A. L. Reysenbach, and M. Podar. 2016. Genomics-informed isolation and characterization of a symbiotic Nanoarchaeota system from a terrestrial geothermal environment. *Nature Communications* 7:12115.

The Largest and Smallest Bacteria

Fraser, C. M., J. D. Gocayne, O. White, M. D. Adams, R. A. Clayton, R. D. Fleischmann, C. J. Bult, A. R. Kerlavage, G. Sutton, J. M. Kelley, J. L. Fritchman, J. F., Weidman, K. V. Small, M. Sandusky, J. Fuhrmann, D. Nguyen, T. R. Utterback, D. M Saudek, C. A. Phillips, J. M. Merrick, J. F. Tomb, B. A. Dougherty, K. F. Bott, P. C. Hu, T. S. Lucier, S. N. Peterson, H. O. Smith, C. A. Hutchison, and J. C. Venter. 1995. The minimal gene complement of *Mycoplasma genitalium*. *Science* 270 (5235):397–403.

Levin, P. A., and E. R. Angert. 2015. Small but mighty: Cell size and bacteria. *Cold Spring Harbor Perspectives in Biology* 7 (7):a019216.

Salman, V., J. V. Bailey, and A. Teske. 2013. Phylogenetic and morphologic complexity of giant sulphur bacteria. *Antonie Van Leeuwenhoek* 104 (2):169–86.

Schulz, H. N., T. Brinkhoff, T. G. Ferdelman, M. Hernández Mariné, M., A. Teske, and B. B. Jørgensen. 1999. Dense populations of a giant sulfur bacterium in Namibian shelf sediments. *Science* 284 (5413):493–95.

The Largest Archaeal Cells and Multicellular Structures

Moissl, C., C. Rudolph, and R. Huber. 2002. Natural communities of novel Archaea and Bacteria with a string-of-pearls-like morphology: Molecular analysis of the bacterial partners. *Applied and Environmental Microbiology* 68 (2):933–37.

Muller, F., T. Brissac, N. Le Bris, H. Felbeck, and O. Gros. 2010. First description of giant Archaea (*Thaumarchaeota*) associated with putative bacterial ectosymbionts in a sulfidic marine habitat. *Environmental Microbiology* 12 (8):2371–83.

The Smallest Animals and the Smallest Free-Living Eukaryote

Courties, C., A. Vaquer, M. Troussellier, J. Lautier, M. J. Chrétiennot-Dinet, J. Neveux, C. Machado, H. Claustre. 1994. Smallest eukaryotic organism. *Nature* 370 (6487):255.

Mockford, E. 1997. A new species of *Dicopomorpha* (Hymenoptera: Mymaridae) with diminutive, apterous males. *Annals of the Entomological Society of America* 90 (2):115–20.

Mohrbeck, I., P. Martinez Arbizu, and T. Glatzel. 2010. Tantulocarida (Crustacea) from the Southern Ocean deep sea, and the description of three new species of *Tantulacus*. In R. Huys et al., *Systematic Parasitology* 77:131–51.

Giant Viruses and Emergent Biological Properties of Viruses

Abergel, C., M. Legendre, and J. M. Claverie. 2015. The rapidly expanding universe of giant viruses: Mimivirus, Pandoravirus, Pithovirus and Mollivirus. *FEMS Microbiology Reviews* 39 (6):779–96.

Erez, Z., I. Steinberger-Levy, M. Shamir, S. Doron, A. Stokar-Avihail, Y. Peleg, S. Melamed, A. Leavitt, A. Savidor, S. Albeck, G. Amitai, and R. Sorek. 2017. Communication between viruses guides lysis-lysogeny decisions. *Nature* 541 (7638):488–93.

Legendre, M., J. Bartoli, L. Shmakova, S. Jeudy, K. Labadie, A. Adrait, M. Lescot, O. Poirot, L. Bertaux, C. Bruley, Y. Couté, E. Rivkina, C. Abergel, and J. M. Claverie. 2014. Thirty-thousand-year-old distant relative of giant icosahedral DNA viruses with a pandoravirus morphology. *Proceedings of the National Academy of Sciences USA* 111 (11):4274–79.
    The discovery of the exceptionally large *Pithovirus*, which infects amoebas.

Long-Term Survivability of Microbes on Earth

Anderson, A. W., H. C. Nordon, R. F. Cain, G. Parrish, and D. Duggan. 1956. Studies on a radio-resistant micrococcus. I. Isolation, morphology, cultural characteristics, and resistance to gamma radiation. *Food Technology* 10:575–77.
    This was the first isolation of *Deinococcus radiodurans*. The discovery happened serendipitously while the authors were testing a method for sterilizing food with high doses of gamma radiation. The bacterium that would be named *Deinococcus radiodurans* survived inside, and later spoiled, a tin can of meat exposed to the radiation.

Colwell, F. S., and S. D'Hondt. 2013. Nature and extent of the deep biosphere. *Reviews in Mineralogy and Geochemistry* 75 (1):547–74.

Leuko, S., L. Rothschild, and B. P. Burns. 2010. Halophilic archaea and the search for extinct and extant Life on Mars. *Journal of Cosmology* 5:940–50.

Orcutt, B. N., D. E. Larowe, J. F. Biddle, F. S. Colwell, B. T. Glazer, B. K. Reese, J. B. Kirkpatrick, L. L. Lapham, H. J. Mills, J. B. Sylvan, S. D. Wankel, and C. G. Wheat. 2013. Microbial activity in the marine deep biosphere: Progress and prospects. *Frontiers in Microbiology* 4:189.

Sankaranarayanan, K., T. K. Lowenstein, M. N. Timofeeff, B. A. Schubert, and J. K. Lum. 2014. Characterization of ancient DNA supports long-term survival of Haloarchaea. *Astrobiology* 14 (7):553–60.

Setlow, P. 2007. I will survive: DNA protection in bacterial spores. *Trends in Microbiology* 15 (4):172–80.

Slade, D., and M. Radman. 2011. Oxidative stress resistance in *Deinococcus radiodurans*. *Microbiology and Molecular Biology Reviews* 75 (1):133–91.

Jupiter's Moon Europa and Its Astrobiological Features

Chyba, C. F., and C. B. Phillips. 2001. Possible ecosystems and the search for life on Europa. *Proceedings of the National Academy of Sciences USA* 98 (3): 801–04.

McCarthy, C., and Cooper, R. F. 2016. Tidal dissipation in creeping ice and the thermal evolution of Europa. *Earth and Planetary Science Letters* 443:185–94.

McCord, T. B., G. B. Hansen, D. L. Matson, T. V. Johnson, J. K. Crowley, F. P. Fanale, R. W. Carlson, W. D. Smythe, P. D. Martin, C. A. Hibbitts, J. C. Granahan, and A. Ocampo. 1999. Hydrated salt minerals on Europa's surface from the Galileo near-infrared mapping spectrometer (NIMS) investigation. *Journal of Geophysical Research: Planets* 104 (E5):11827–51.

Phillips, C. B., and R. T. Pappalardo. 2014. Europa Clipper Mission concept: Exploring Jupiter's ocean moon. *Earth & Space Science News* 95 (20): 165–67.

Roth, L., J. Saur, K. D. Retherford, D. F. Strobel, P. D. Feldman, M. A. McGrath, F. Nimmo. 2013. Transient water vapor at Europa's South Pole. *Science* 343 (6167):171–74.

Sagan, C. 1971. The Solar System beyond Mars: An exobiological survey. *Space Science Reviews* 11 (6):827–866.

Sparks, W. B., K. P. Hand, M. A. McGrath, E. Bergeron, M. Cracraft, and S. E. Deustua. 2016. Probing for evidence of plumes on Europa with HST / STIS. *Astrophysical Journal* 829 (2): 121–42.

Capabilities of the James Webb Space Telescope

Jakobsen, P., and P. Jensen. 2008. James Webb Space Telescope: A bigger and better time machine. *ESA Bulletin* 133: 32–40.

Big-Bang Cosmology: The Age and Large-Scale Structure of the Universe

Boylan-Kolchin, M., V. Springel, S. D. White, A. Jenkins, and G. Lemson. 2009. Resolving cosmic structure formation with the Millennium-II Simulation. *Monthly Notices of the Royal Astronomical Society* 398 (3):1150–64.
  Simulations of matter distribution in the universe and the cosmic web

Conselice, C. J., A. Wikinson, K. Duncan, and A. Mortlock. 2016. The evolution of galaxy number density at *z*<8 and its implications. *Astrophysical Journal* 830 (2): 83–100.
  This study predicts that there are two trillion galaxies in the universe.

Hinshaw, G., J. L. Weiland, R. S. Hill, N. Odegard, D. Larson, C. L. Bennett, J. Dunkley, B. Gold, M. R. Greason, N. Jarosik, E. Komatsu, M. R. Nolta, L. Page, D. N. Spergel, E. Wollack, M. Halpern, A. Kogut, M. Limon, S. S. Meyer, G. S. Tucker, and E. L. Wright. 2009. Five-year Wilkinson Microwave Anisotropy Probe (WMAP) observations: Data processing, sky maps, and basic results. *Astrophysical Journal Supplement Series* 180 (2):225–45.
  Data from this project produced an all-sky map of the cosmic microwave background (CMB), sometimes called a "baby picture" of the universe. CMB is considered the first recordable light in the cosmos. Measurements by the WMAP found that the universe is 13.77 billion years old. Subsequent measurements from the Planck space observatory gave a higher resolution view of the CMB, dating the universe at 13.82 billion years old.

How to Photograph Microbes

Cybulski, J. S., J. Clements, and M. Prakash. 2014. Foldscope: Origami-based paper microscope. *PLoS One* 9 (6):e98781.

Diamantis, A., E. Magiorkinis, and G. Androutsos. 2009. Alfred Françcois Donné (1801–78): A pioneer of microscopy, microbiology and haematology. *Journal of Medical Biography* 17 (2): 81–87.

Masters, Barry R. 2010. The development of fluorescence microscopy. In *Encyclopedia of Life Sciences* (ELS). Chichester, UK: John Wiley & Sons.

Overney, N., and G. Overney. 2011. *The History of Photomicrography,* 3rd ed. http://microscopy-uk.org.uk/mag/artmar10/history_photomicrography_ed3.pdf

Sydor, A. M., K. J. Czymmek, E. M. Puchner, and V. Mennella. 2015. Super-resolution microscopy: From single molecules to supramolecular assemblies. *Trends in Cell Biology* 25 (12):730–48.

ACKNOWLEDGMENTS

Many people and organizations have been essential supporters of this project. We thank Janice Audet, Executive Editor for Life Sciences at Harvard University Press, for invaluable guidance throughout the entire process of writing this book. We also thank Jane Pickering and Janis Sacco, directors of the Harvard Museums of Science and Culture, for supporting the Microbial Life exhibition at the Harvard Museum of Natural History, which was developed alongside the writing of this book. We are grateful to the anonymous reviewers of the manuscript for their feedback, and to Kate Brick and Lisa Roberts of Harvard University Press for their outstanding contributions in editing and design. And we thank Elio Schaechter for reading the manuscript and for sharpening both the text and the images through his commentary.

We thank fellow members of the Kolter Lab for their support, including Lori Shapiro, Nick Lyons, Gleb Pischany, Jorge Rocha, Ben Niu, Einat Segev, and Anna Depetris. We are particularly indebted to them for their deep and thoughtful discussions on this project during our annual retreat, the "Maine Event" at Howells House in Kittery Point, Maine. Their input was critical in providing clear direction for the project.

We appreciate several others who helped in the gathering of photographs and information. Thanks to geologist Len Eisenberg for providing maps and guidance in locating the stromatolite fossils in Capitol Reef National Park, and to Hannah Bonner, National Park Service ranger at Capitol Reef. Gilles van Wezel was an excellent host in the Netherlands and liaison to the history of Dutch microbiology. Also in the Netherlands, several photographs in this book were captured within *Micropia,* the world's first microbe zoo at the Royal Artis Zoo in Amsterdam. We also thank Andrew Knoll for discussions on stromatolite fossils and Douglas Eveleigh for discussions on the discovery of streptomycin at Rutgers University.

We thank the Harvard University Herbaria and Libraries for access to photograph Roland Thaxter's original drawings of myxobacteria and Cohn and Koch's drawings of *Bacillus.* Thanks also to Jean-François Gauvin of the Harvard Collection of Historical Scientific Instruments, who provided access to their historic microscope collections.

Jessica Mark Welch of the Marine Biological Laboratory at Woods Hole, Massachusetts, and Gary Borisy of the Forsyth Institute provided the image of human dental plaque microbiome shown in Chapter 1. Several photographs in this book were taken in Ben Wolfe's microbial foods lab at Tufts University. Ben Wolfe also contributed a few images to Chapter 6. Two former members of the Kolter Laboratory, Hera Vlamakis and Jordi van Gestel, provided one image each to Chapter 4.

We thank David Goodsell and the Research Collaboratory for Structural Bioinformatics (RCSB) *Molecule of the Month* series for the paintings seen in several chapters that add a size scale below the microbial world. And we thank the National Aeronautics and Space Administration (NASA) and its branches and affiliates for covering size scales above the microbial and human scales!

Scanning electron microscopy seen in this book was conducted at the Center for Nanoscale Systems (CNS) at Harvard University. Thank you to CNS staff scientists Tim Cavanaugh, Adam Gram, and Carolyn Marks for their help and maintenance of CNS imaging equipment.

From Scott Chimileski

I dedicate my portion of the work on this book to two of my relatives who have passed away: to my great uncle, René Pauli, whose nature photography prints inspired me as a child and to this day, and to his sister, my grandmother, Ines (Schatzi) Freuler, who supported René and also nurtured my creativity. I thank my parents, Ken Chimileski and Barbara Freuler Chimileski, for taking me to experience wild places as a child. And I thank my sister, Lindsay Chimileski, and my brothers, Andrew and Brock Chimileski, for continuing to explore wilderness with me today. I am grateful to my family in general and to Andrew in particular for his interest in this project since day one.

I would also like to thank the US National Park Service and the rangers and staff at Capitol Reef National Park, Yellowstone National Park, and Glacier National Park, who keep these inspiring places wild and safe. My photography trips to these parks in 2016 coincided with the one-hundred-year anniversary of the National Park Service, and I sincerely hope there will be many more anniversaries of this great organization in the future.

Finally, I cannot express my appreciation enough for my coauthor and mentor Roberto Kolter, for all that he has done to inspire me, and for all that he has done to inspire so many other young microbiologists since first opening his laboratory in 1983.

From Roberto Kolter

This book project had been brewing in my mind for more years than I wish to acknowledge. But it took a stroke of good fortune for it to come to be: Scott Chimileski applied to become a research fellow in the Kolter Lab. Once we joined forces, it was clear that it would become a reality. It was his remarkable artistic talent that brought this book to fruition, and for that and much more, I remain forever grateful. Many thanks also to Jon and Andrea Clardy, who provided me with their very enthusiastic reaction to and support of the work. My most heartfelt thanks to my wife, Mechas Zambrano, who has steadfastly encouraged me to pursue my oftentimes farfetched ideas. Her ability to support and particularly to challenge my thinking was instrumental during every step of making this book a reality.

IMAGE INFORMATION AND CREDITS

Eighty percent of the illustrations in this book are images that we have produced within the past two years to illustrate the organisms and concepts we discuss and to guide the stories we tell. In many cases we captured these images in the laboratory. But we also traveled to find and photograph "wild" microbes and natural microbial ecosystems. Our exploration took us to many local sites in New England but also across the world, to the birthplace of microbiology in the Netherlands, the White Cliffs of Dover in the United Kingdom, Great Salt Lake in Utah, and Yellowstone and other national parks of the United States. The remaining 20 percent of images are a curated combination of historic and scientific illustrations and other works of art, imagery of Earth and celestial objects from NASA and affiliate government agencies, and images brought in from scientific papers or from other scientists specializing in particular fields or imaging techniques. Unless otherwise noted, all photographs, light microscopy, three-dimensional models, paintings, and scanning electron micrographs are by Scott Chimileski. Scanning electron microscopy was conducted at the Center for Nanoscale Systems at Harvard University.

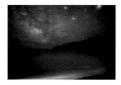

p. ii Milky Way galaxy over Grand Prismatic Spring.
Yellowstone National Park, Wyoming. Photograph, 2016.

p. vi Dew drop on a blade of grass.
Arnold Arboretum, Harvard University, Jamaica Plain, Massachusetts.
Photograph, 2017.

p. viii Mammoth Hot Springs.
Yellowstone National Park, Wyoming. Photograph, 2016.

1. From an Ancient Chalk Graveyard

p. xiv The White Cliffs of Dover, United Kingdom.
Photograph, 2016.

p. 2 (left) Crumbles of chalk from the White Cliffs.
Dover, United Kingdom. Photograph, 2016.

p. 2 (right) A replica of Antoni van Leeuwenhoek's microscope.
Collection of Historical Instruments, Harvard University.
Photograph, 2016.

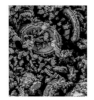

p. 3 Mysterious disc-shaped objects.
Scanning electron micrograph. Photograph, 2016

pp. 4–5 Light blue and green patches in the North Atlantic Ocean beneath the clouds.
Satellite image, June 1998. SeaWiFS Project/NASA/Goddard Space Flight Cente/
ORBIMAGE

p. 6 Brilliant patterns swirling in the South Atlantic Ocean.
Satellite image, November 2015. NASA/Ocean Biology Processing Group/Goddard
Space Flight Center.

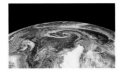

p. 8 The route of the first transatlantic telegraph cable.
Map from *Frank Leslie's Illustrated Newspaper,* August 21, 1858.

p. 9 (left) Round coccolith structures.
Scanning electron micrograph, 2016.

p. 9 (right) The coccolithophore *Emiliania huxleyi.*
Scanning electron micrograph, 2016.

p. 10 (left) A full season of ocean chlorophyll concentrations.
Satellite image, June to September 2012. NASA/Suomi NPP/Norman Kuring.

p. 10 (right)   A phytoplankton bloom in the North Sea.
Satellite image, October 1997. SeaWiFS Project/NASA Goddard Space Flight Center/ORBIMAGE.

p. 12 (above)   Map of Delft, the Netherlands.
Joan Blaeu, *Toonneel der steden van de Vereenighde Nederlanden, met hare beschrijvingen* (Amsterdam: J. Blaeu, 1652).

p. 12 (below)   *View of Delft,* Johannes Vermeer, 1660–1661.
Oil on canvas. Mauritshuis Royal Picture Gallery, the Hague, the Netherlands.

p. 13   Bacterial cells as seen through Leeuwenhoek's microscope.
Antoni van Leeuwenhoek, Letter No. 39, to Francis Aston, September 17, 1683. The Royal Society, London (MS 1898.L1.69). Enlarged from the engravings published in *Arcana Naturae Detecta,* 1695.

p. 14   The tree diagram from *On the Origin of Species.*
Charles Darwin, *On the Origin of Species.* (London: John Murray, 1859).

p. 15   The three-domain tree of life.
Metal sculpture commissioned by Roberto Kolter. Photograph, 2017.

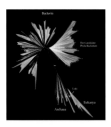

p. 17 A new tree of life.
Modified from L. A. Hug et al. A new view of the tree of life. *Nature Microbiology* 1:16048 (2016). Figure 1. (CC BY 4.0)

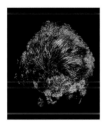

p. 18 The hedgehog microbiome. This fluorescence microscopy image was produced using the combinatorial labeling and spectral imaging fluorescence in situ hybridization method (CLASI-FISH).
Micrograph by Jessica Mark Welch of the Marine Biological Lab of Woods Hole, Massachusetts, in collaboration with Gary Borisy and Floyd Dewhirst of the Forsyth Institute.

p. 20 An artistic rendition of the complex and crowded molecular realm inside an *E. coli* cell.
Watercolor painting of *Escherichia coli*. David S. Goodsell, the Scripps Research Institute. © David S. Goodsell, 1999.

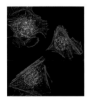

p. 21 The eukaryotic cell.
Micrograph by Dylan Burnette and Jennifer Lippincott-Schwartz, Eunice Kennedy Shriver National Institute of Child Health and Human Development, National Institutes of Health.

p. 22 Aerial view of Delft.
Delft, the Netherlands. Photograph, 2015.

p. 23 (left)   The canal at the site of Antoni van Leeuwenhoek's house. Delft, the Netherlands. Photograph, 2015.

p. 23 (right)   A miniature ecosystem grows at the edge of the canal. Delft, the Netherlands. Photograph, 2015.

p. 25   Mud Volcano Area, near Dragon's Mouth Spring. Yellowstone National Park, Wyoming. Photograph, 2016.

p. 26   The canals of Delft filled with duckweed. Delft, the Netherlands. Photograph, 2016.

p. 27   Every frond of duckweed is a microbial ecosystem. Delft, the Netherlands. Photograph, 2016.

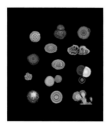

p. 28   A variety of bacterial and fungal species form macroscopic colonies. Photographs, 2016.

2. To the Heartbeat of Earth

p. 30 Grand Prismatic Spring, Yellowstone National Park, Wyoming.
Light painting by Andrew Chimileski. Photograph, 2016.

p. 32 Grand Prismatic Spring and Excelsior Geyser Crater.
Yellowstone National Park, Wyoming. Aerial photograph, Jim Peaco/Yellowstone
Historic Photo Collection/Yellowstone National Park/U.S. National Park Service.

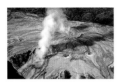

p. 33 Tourists walk over the orange microbial mats at Grand Prismatic Spring.
Yellowstone National Park, Wyoming. Photograph, 2016.

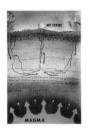

p. 34 Vintage diagram of geothermal features.
U.S. National Park Service, Yellowstone Historic Photo Collection, Yellowstone
National Park, Wyoming, 1978.

p. 35 Carina Nebula.
Image captured by the Hubble Space Telescope. NASA/European Space Agency/
M. Livio, The Hubble Heritage Team and the Hubble 20th Anniversary Team (STScI),
2010. heic1007e (CC-BY-4.0).

p. 36 The supernova SN 1987.
Composite image from the Hubble Space Telescope. ALMA (ESO/NAOJ/NRAO)/
A. Angelich, 2014. Visible light image: the NASA/ESA Hubble Space Telescope.
X-ray image: The NASA Chandra X-Ray Observatory, eso1401a (CC-BY-4.0)

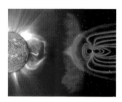

p. 37 The magnetosphere shields Earth from the ever-present solar wind.
© SOHO/LASCO/EIT (ESA & NASA). Illustration by Steele Hill, NASA, 2002.

p. 39 (above) Mars, without its magnetosphere.
Composite image of 1,000 red and violet images captured by the spacecraft *Viking Orbiter 1*. NASA Jet Propulsion Laboratory and the US Geological Survey. NASA/JPL-Caltech/USGS, 1998.

p. 39 (below) Aurora and sunrise over northern Europe.
Photograph, Expedition 43 (International Space Station), March 2015. NASA/Johnson Space Center.

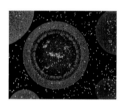

p. 43 A scientifically accurate depiction of protocells, based on work from laboratories attempting to recreate protocells from molecules today.
Illustration by Janet Iwasa, in collaboration with Jack Szostak. Szostak Laboratory, Harvard Medical School and Massachusetts General Hospital, and the Current Science and Technology Team at the Museum of Science, Boston.

p. 45 The structure of a bacterial ribosome.
Watercolor painting by David S. Goodsell. *RCSB PDB Molecule of the Month*, Scripps Research Institute and the Research Collaboratory for Structural Bioinformatics (RCSB) Protein Data Bank (PDB), January 2010. (CC-BY-4.0)

p. 47 A vision of the early Earth.
Foreground: Sapphire Pool, Yellowstone National Park, Wyoming, 1902. Detroit Photographic Company. Background: "Pluto at High Noon" (artist's concept) NASA/Jet Propulsion Laboratory/Southwest Research Institute/Alex Parker, PIA19682. Composite image, 2016.

p. 48 Glacier National Park, northern Montana, satellite view.
NASA Landsat 7 image created by Jesse Allen, Earth Observatory, July 7, 2001 using data obtained courtesy of the University of Maryland's Global Land Cover Facility.

p. 49 (above) An outcrop of the Altyn formation along the Going-to-the-Sun Road.
Glacier National Park, Montana. Photograph, 2016.

p. 49 (below) Texture of layered, dome-shaped stromatolite fossils.
Glacier National Park, Montana. Photograph, 2016.

p. 50 Sandstone formations within Capitol Reef National Park.
Photograph by René Pauli.

pp. 52–53 A large outcrop of stromatolite fossils within Capitol Reef National Park.
Photograph, 2016.

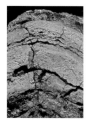

p. 54 Cross section of stromatolite fossils from Capitol Reef National Park.
From the fossil collection of Len Eisenberg. Photograph, 2016.

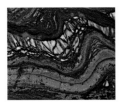

p. 55   Polished cross section of banded iron formations.
Pilbara region, Western Australia. Photograph, 2016.

p. 56 (above)   Octopus Spring.
Yellowstone National Park, Wyoming. Photograph, 2016.

p. 56 (below)   A thermal stream lined with microbial mats, Octopus Spring.
Yellowstone National Park, Wyoming. Photograph, 2016.

p. 57   A nursery of growing microbialite structures.
Yellowstone National Park, Wyoming. Photograph, 2016.

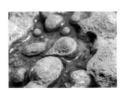

p. 58 (above)   Close-up of tiny pearl-shaped microbial structures.
Yellowstone National Park, Wyoming. Photograph, 2016.

p. 58 (below)   Microbial mat, side view.
Yellowstone National Park, Wyoming. Photograph, 2016.

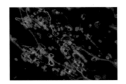

p. 59   A community of thermophilic microbes.
Fluorescence microscopy with UV excitation. Micrograph, 2016.

p. 60   The upper end of Great Salt Lake, Utah.
    Satellite image. NASA Earth Observatory.

pp. 62–63   Microbialites grow in salty waters.
    Buffalo Point on Antelope Island, Great Salt Lake, Utah. Photograph, 2016.

p. 64   Green cyanobacteria form layers within the Great Salt Lake microbialites.
    Buffalo Point on Antelope Island, Great Salt Lake, Utah. Photograph, 2016.

p. 65   Water evaporates in a corner along a beach.
    Farmington Bay, Antelope Island, Great Salt Lake, Utah. Photograph, 2016.

p. 66   A biofilm of red halophilic microbes.
    Farmington Bay, Antelope Island, Great Salt Lake, Utah. Photograph, 2016.

p. 67 (above)   Texture of the mud biofilm with a streak of yellow algae.
    Farmington Bay, Antelope Island, Great Salt Lake, Utah. Photograph, 2016.

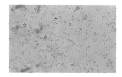

p. 67 (below) The natural pigments of Great Salt Lake microbes.
   Micrograph, 2016.

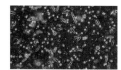

pp. 68–69 Green diatoms and other halophilic microbes from Great Salt Lake mud.
   Micrograph, 2016.

p. 70 Great Salt Lake mud biofilm as seen with electron microscopy.
   Direct preparation of mud sample, undiluted. Scanning electron micrograph, 2016.

p. 71 Lamar Valley at sunset.
   Yellowstone National Park, Wyoming. Photograph, 2016.

p. 72 American bison.
   Lamar Valley, Yellowstone National Park, Wyoming. Photograph, 2016.

3. Under Celia Thaxter's Garden

p. 74   Celia and Eliot Thaxter in *The Garden in Its Glory* by Childe Hassam, 1892.
Watercolor on paper. Smithsonian American Art Museum. Gift of John Gellatly.

p. 76   Appledore House on the Isles of Shoals, 1890s.
Postcard, *Celia Thaxter Cottage, Isle-of-Shoals, N.H.*, Detroit Publishing Company, 1901.

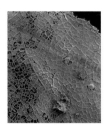

p. 78   A leaf on the forest floor.
Concord, Massachusetts, near Walden Pond. Scanning electron micrograph, 2016.

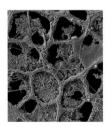

p. 79   Fungal hyphae, bacteria and spores among the remaining veins of the leaf.
Concord, Massachusetts, near Walden Pond. Scanning electron micrograph, 2016.

p. 80   Soil.
Kittery Point, Maine. Photograph, 2016.

p. 81 White fungal hyphae.
   Kittery Point, Maine. Photograph, 2016.

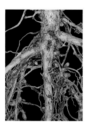

p. 83 Root nodules of a clover from the genus *Trifolium*.
   New Haven, Connecticut. Photograph, 2016.

p. 84 The pink hue of root nodules.
   New Haven, Connecticut. Photograph, 2016.

p. 86 A pill bug feeds on decaying plant material.
   Concord, Massachusetts, near Walden Pond. Photograph, 2016.

p. 88 Trail into a sunlit forest.
   Kittery Point, Maine. Photograph, 2016.

p. 90 Colonies formed by diverse microbes living on a fern leaf.
  Kittery Point, Maine. Photograph, 2016.

p. 91 Microbial biodiversity on an oak leaf.
  Kittery Point, Maine. Photograph, 2016.

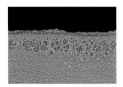

p. 92 (above) The forest floor.
  Kittery Point, Maine. Photograph, 2016.

p. 92 (below) A colony of lichen on a decaying log.
  Kittery Point, Maine. Photograph, 2016.

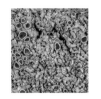

p. 93 Cross section of lichen.
  Micrograph, 2016.

p. 94 Symbiosis between fungal hyphae and algal cells inside of a lichen.
  Scanning electron micrograph, 2016.

p. 95 Mushroom caps from the genus *Mycena*.
    Concord, Massachusetts, near Walden Pond. Photograph, 2016.

p. 96 (above) The swamp ecosystem.
    Kittery Point, Maine. Photograph, 2016.

p. 96 (below) Water droplets on moss.
    Kittery Point, Maine. Photograph, 2016.

p. 97 Microbes within the moss.
    Scanning electron micrograph, 2016.

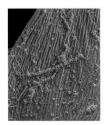

p. 98 Microbial life adheres to moss leaves.
    Scanning electron micrograph, 2016.

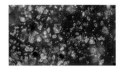

p. 99 Moss microbes scattered among grains of sand.
    Dark-field microscopy. Micrograph, 2016.

p. 100 (above, left) Rotifer from the genus *Philodina* (imaged live).
Micrograph, 2016.

p. 100 (above, right) Tardigrade, *Hypsibius dujardini* (imaged live).
Micrograph, 2016.

p. 100 (below) Cryptobiosis in tardigrades, sp. *Ramazzottius varieornatus.*
Image modified from T. Hashimoto et al. 2016. Extremotolerant tardigrade genome
and improved radiotolerance of human cultured cells by tardigrade-unique protein.
Figure 1 (a, b), *Nature Communications* 7:12808. (CC BY-NC-SA 4.0)

p. 101 The salt marsh at the end of the trail.
Kittery Point, Maine. Photograph, 2016.

p. 102 Spongy green patches within the salt marsh.
Kittery Point, Maine. Photograph, 2016.

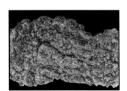

p. 103 A slice through the spongy marsh microbial mat.
Kittery Point, Maine. Photograph, 2016.

p. 104 Red autofluorescent cyanobacteria within salt marsh microbial mats.
Fluorescence microscopy with yellow excitation. Micrograph, 2016.

p. 105   A nematode meanders through the matrix of the microbial mat.
Scanning electron micrograph, 2016.

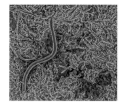

p. 106   Tidal pools on the beach.
Kittery Point, Maine. Photograph, 2016.

p. 107   Smooth rocks and algae line each tidal pool.
Kittery Point, Maine. Photograph, 2016.

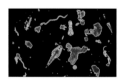

p. 108 (above)   A zoo of micro-animals and algae in every tidal pool.
Micrograph, 2016.

p. 108 (below)   The autofluorescent pigments of tidal pool microbes.
Fluorescence microscopy with yellow excitation. Micrograph, 2016.

p. 109   *Paramecium caudatum* (imaged live).
Micrograph, 2016.

p. 110   The microbial aquascape in the crevices of a submerged tidal pool rock.
  Scanning electron micrograph, 2016.

p. 111   The perforated glass outer shell of diatom cells.
  Scanning electron micrograph, 2016.

p. 113   Colonies develop from bacteria left by fingerprints on a solid growth medium.
  Photograph, 2016.

4. INTELLIGENT SLIME

p. 115   Harvard Medical School and surrounding buildings.
  Longwood Medical Area, Boston, Massachusetts. Photograph, 2016.

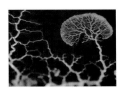

p. 116   Petri plates stacked in a microbiology laboratory.
  Kolter Lab, Harvard Medical School, Boston, Massachusetts. Photograph, 2016.

p. 117   The slime mold *Physarum polycephalum*.
  Photograph, 2016.

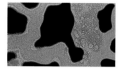

pp. 118–119   A contiguous cytoplasm with many free nuclei pumps through the web of tubes.
  Micrograph, 2016.

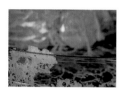

p. 120   *Physarum* escapes from its petri plate.
  Photograph, 2016.

p. 124   A first look at the biology of *Bacillus* species, including drawings by Ferdinand Cohn and Robert Koch in separate works on similar species, both published in 1877.
  Photograph of original plate accessed at the Harvard University Herbaria & Libraries. Figures 1–7 by Dr. Ferdinand (Julius) Cohn; Figures 8–11 by Dr. Robert (Heinrich Hermann) Koch. In Ferdinand Cohn, ed., *Beiträge zur Biologie der Pflanzen*, vol. 2, pt. 2, plate 11 (Breslau: J. U. Kern's Verlag (Max Müller), 1877), 277–310.

p. 125   A floating pellicle biofilm of *Bacillus subtilis.*
  Photograph, 2016.

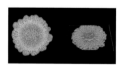

p. 126 (above)   The three-dimensional structure of *Bacillus subtilis* colony biofilms.
  Photograph and 3D model, 2016.

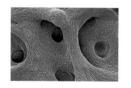

p. 126 (below).   Cells within *Bacillus* biofilms linked by a common extracellular matrix.
  Scanning electron micrograph by Roberto Kolter and Steve Minsky.

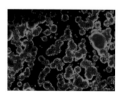

p. 127   Water channels allow flow throughout a submerged biofilm.
  Fluorescence microscopy with blue excitation. Micrograph, 2016.

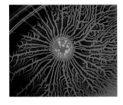

p. 128   *Bacillus subtilis* swarms as one.
Petri plate prepared by Nick Lyons. Photograph, 2016.

p. 129   *Pseudomonas aeruginosa* develops as a colony biofilm.
Photographs, 2016.

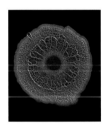

p. 130   The ornate, wrinkled structure of a mature *Pseudomonas aeruginosa* biofilm.
Composite photograph, 2016.

p. 131   Extracellular matrix and pili connect cells in a *Pseudomonas aeruginosa* biofilm.
Scanning electron micrograph, 2016.

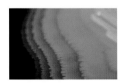

p. 132   *Proteus mirabilis* forms terraced patterns as it swarms across a surface.
Photograph, 2016.

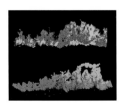

p. 133   Different functions occur at different places inside a *Bacillus subtilis* biofilm.
3D model *(above)* showing the cross-section through a *Bacillus* biofilm produced in 2015. Micrographs *(below)* by Hera Vlamakis, Kolter lab, Microbial Sciences Initiative, Harvard Medical School. See also Vlamakis et al. 2008. Control of cell fate by the formation of an architecturally complex bacterial community. *Genes and Development* 22 (7):945–53.

p. 135 Sliding motility: another example of division of labor in *Bacillus* communities.
Photograph (above) and micrograph (below) by Jordi van Gestel, Kolter lab, Microbial Sciences Initiative, Harvard Medical School. See also J. van Gestel, H. Vlamakis, and R. Kolter, From cell differentiation to cell collectives: *Bacillus subtilis* uses division of labor to migrate. PloS Biology 13 (4):e1002141

p. 136 (above) The myxobacterium *Myxococcus xanthus* spreads across a surface over several days.
Photographs, 2015.

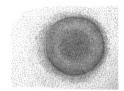

p. 136 (below) Multicellular movements of a species from the genus *Myxococcus* as drawn by Roland Thaxter, 1892.
Photograph of original drawings accessed at the Harvard University Herbaria & Libraries. Roland Thaxter, On the Myxobacteriaceae, a new order of Schizomycetes. Contributions from the Cryptogamic Laboratory of Harvard University. *Botanical Gazette* 17 (12):389–406 (1892), plate 25, fig. 37.

p. 138 Swarms of different *Bacillus subtilis* strains either merge or form boundaries.
Petri plate prepared by Nick Lyons, Kolter lab, Microbial Sciences Initiative, Harvard Medical School. Composite photograph, 2016.

p. 139 The multitude of microbial cell morphologies.
Adapted from figure 1 in D. T. Kysela, A. M. Randich, P. D. Caccamo, and Y. V. Brun. 2016. Diversity takes shape: Understanding the mechanistic and adaptive basis of bacterial morphology. *PLoS Biology* 14(10):e1002565. (CC-BY-4.0)

p. 141 (above) Artistic concept of individual *Prochlorococcus* cells.
Watercolor painting, 2017.

p. 141 (below) Cyanophage viruses that infect *Prochlorococcus* cells.
Features of bacteriophage anatomy seen here are the capsid head, contractile tail (sheath) and tail fibers used to attach to bacterial cells. Adapted from fig. 2 in M. B. Sullivan et al. 2005. Three *Prochlorococcus* cyanophage genomes: signature features and ecological interpretations. *PLoS Biology* 3(5):e144. (CC-BY-4.0)

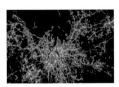

p. 144 (above) A *Physarum* network as seen from above.
Photograph, 2016.

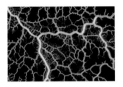

p. 144 (below) The networked roads and lights of Boston, Massachusetts as seen from a satellite.
Photograph captured from the International Space Station. Earth Science and Remote Sensing Unit, Johnson Space Center, NASA.

p. 145 Ernst Haeckel's illustration of twenty different macroscopic slime molds.
Ernst (Heinrich Philipp) Haeckel, *Kunstformen der Natur [Art Forms in Nature]* (Leipzig und Wien: Verlag des Bibliographischen Instituts, 1904), plate 93: Mycetozoa.

5. TALES OF SYMBIOSIS

p. 146 A soldier leaf-cutter ant defends smaller worker ants.
Yale Peabody Museum of Natural History, New Haven, Connecticut. Photograph, 2016.

p. 148 Map of Llewelyn Williams's Botanical Field Expedition in 1929.
From Llewelyn Williams, *Woods of Northeastern Peru* (Chicago: Field Museum of Natural History, 1936). Field Museum of Natural History, Chicago. (CC BY-NC-SA 3.0) Williams's adventures observing ants were reported in "Jungle Surgery," an article in *Time* magazine, April 28, 1930, published upon his return to Chicago.

p. 149 (above) A worker leaf-cutter ant cuts sections of leaf.
Micropia, Royal Artis Zoo, Amsterdam, the Netherlands. Photograph, 2016.

p. 149 (below) Workers forage from the forest floor all way up to the rainforest canopy.
Micropia, Royal Artis Zoo, Amsterdam, the Netherlands. Photograph, 2016.

p. 150 Scavenged leaves are brought into an underground network of chambers.
Illustration by Thomas Belt in *The Naturalist in Nicaragua* (London, J.M. Dent & Sons, Ltd.; New York, E.P. Dutton & Co., 1911). Smithsonian Libraries.

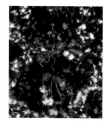

p. 152 Leaf-cutter ants farm a crop of fungus.
Micropia, Royal Artis Zoo, Amsterdam, the Netherlands. Photograph, 2016.

p. 153 A smaller subtype of ants specializes in tending the fungal gardens.
Micropia, Royal Artis Zoo, Amsterdam, the Netherlands. Photograph, 2016.

p. 156 Albert Schatz's data from "Experiment no. 11: Antagonistic Actinomycetes," August 23, 1943.
Notebook, Special Collections and University Archives, Rutgers University Libraries.

p. 157 The phenomenon of antibiosis.
Photograph, 2015.

p. 158 Colonies of *Streptomyces hygroscopicus*.
Photograph, 2016.

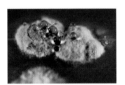

p. 159 (above) *Streptomyces roseosporus* seen with droplets of antibiotic.
Photograph, 2016.

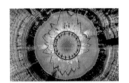

p. 159 (below) A mixed colony of *Streptomyces coelicolor* and *Amycolatopsis* sp. AA4.
Photograph, 2016.

p. 162 The molecule actinomycin intercalates within DNA, blocking RNA transcription.
Watercolor painting by David S. Goodsell. *RCSB PDB Molecule of the Month*, Scripps
Research Institute and the Research Collaboratory for Structural Bioinformatics
(RCSB) Protein Data Bank (PDB), April 2013. (CC-BY-4.0)

p. 163 (left) A monstrous anglerfish with dangling lure.
Redrawn from Plate CXXI, Figure 408 in *Oceanic Ichthyology: A Treatise on the Deep-
Sea and Pelagic Fishes of the World,* by George Brown Goode and Tarleton H. Bean
(Cambridge, MA: Printed for the Museum, John Wilson and Son, 1896), p. 496, a work
of the Harvard Museum of Comparative Zoology based on deep sea dredging expedi-
tions from 1877 to 1880 under the supervision of Alexander Agassiz. Original drawing
from Robert Collett, On a new Pediculate Fish from the Sea off Madeira, *Proceedings
of the Scientific Meetings of the Zoological Society of London for the year 1886* (London:
Messrs. Longmans, Green, and Co., 1886) *Linophryne lucifer*, Plate XV, p. 138.

p. 163 (right) Flashlight fish with cheek-like light organs.
Anomalops palbebratu. Modified from an illustration by G. H. Ford from "Andrew
Garrett's *Fische der Südsee* (*Fishes of the South Seas*)," Book 11, Plate XCI, Figure A,
described and edited by Albert C. G. Gunther in *Journal of the Museum Godeffroy*, vol. 4
(Hamburg: L. Friederichsen & Co., 1873).

p. 164 The movements of a school of flashlight fish.
   Micropia, Royal Artis Zoo, Amsterdam, the Netherlands. Photograph, 2016.

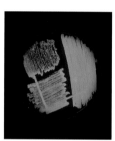

p. 165 Bioluminescent bacteria *Vibrio fisheri* glow inside a petri plate.
   Photograph, 2016.

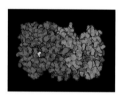

p. 167 The protein bacterial luciferase produces bioluminescence.
   Watercolor painting by David S. Goodsell. *RCSB PDB Molecule of the Month*, Scripps
   Research Institute and the Research Collaboratory for Structural Bioinformatics
   (RCSB) Protein Data Bank (PDB), June 2006. (CC-BY-4.0)

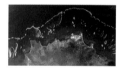

p. 169 (above) The Great Barrier Reef is among the largest living ecosystems.
   Satellite image. NASA/JSC (Johnson Space Center), U.S. Astronaut Kjell Lindgren.

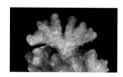

p. 169 (below) A coral from the genus *Pocillopora.*
   Photograph, 2016.

p. 170 Golden zooxanthellae dinoflagellate cells alongside coral nematocysts.
   Micrograph, 2016.

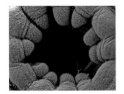

p. 171 (above) The mouth of a *Pocillopora* polyp.
Scanning electron micrograph, 2016.

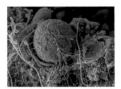

p. 171 (below) A zooxanthella within a symbiosome membrane inside fractured coral tissue.
Scanning electron micrograph, 2016.

p. 173 A commemorative stamp celebrating the discovery of streptomycin.
Issued by Gambia in 1989.

6. ON THE KITCHEN COUNTER

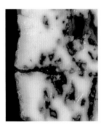

p. 174 Veins of blue mold within the tunnels and pockets inside blue cheese.
Composite photograph, 2016.

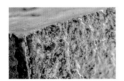

p. 176 The rind of a blue cheese.
Photograph, 2016.

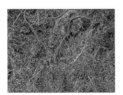

p. 177 (above) The microbial community on a blue cheese rind.
Scanning electron micrograph, 2016.

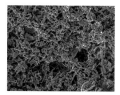

p. 177 (below) Dozens of bacterial species and fungal species grow on the rind.
   Scanning electron micrograph, 2016.

p. 178 Wheels of blue cheese aging in a cheese cave.
   Photograph by Benjamin Wolfe, Tufts University.

p. 179 A cavern of *Penicillium roqueforti* within blue cheese.
   Photograph, 2016.

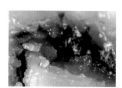

p. 180 Cheese mites live on just about any kind of aged cheese.
   Photograph, 2016.

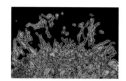

p. 181 *Penicillium* conidiophores with releasing conidia.
   Micrograph, 2017.

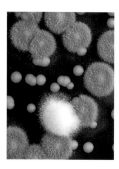

p. 183 A white colony appears among the blue cheese microbes cultivated in the laboratory.
   Photograph by Benjamin Wolfe, Tufts University.

p. 184 Cells of the yeast *Saccharomyces cerevisiae* seen with tiny reproductive buds. Micrograph by Benjamin Wolfe, Tufts University.

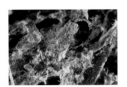

p. 187 *Agaricus bisporus* hyphae form networks of mycelia. Photograph, 2016.

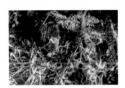

p. 188 Mycelia bud into fruiting bodies. Photograph, 2016.

p. 189 Fruiting bodies grow rapidly into larger buds. Photograph, 2016.

p. 190. Buds at this developmental stage are collected as white button mushrooms. Photograph, 2016.

p. 191 A mature mushroom has formed gills lined with reproductive spores beneath its cap. Photograph, 2016.

p. 192 A kombucha biofilm develops above fermenting black tea.
Photograph, 2016.

p. 193 Cellulose fibers that bind the biofilm together.
Scanning electron micrograph, 2016.

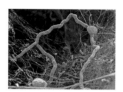

p. 194 (above) Yeast, bacteria, and filamentous mold inside the biofilm.
Scanning electron micrograph, 2016.

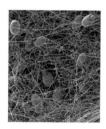

p. 194 (below) Budding yeast cells enmeshed within crisscrossed cellulose fibers.
Scanning electron micrograph, 2016.

p. 195 Microbes cultured from the cabbage leaves used to make fermented kimchi.
Petri plate prepared by Esther Miller in the laboratory of Benjamin Wolfe of
Tufts University. Photograph, 2016.

p. 196 *Aspergillus oryzae* forms filamentous structures.
Photograph, 2016.

p. 197 *Aspergillus oryzae* growing on rice kernels.
Photograph, 2016.

p. 198 Conidia spores release from *Aspergillus oryzae* conidiophores.
Micrograph, 2017.

p. 199 Chocolate is produced from the seeds of the *Theobroma cacao* plant.
Theobroma cacao from F. E. (Franz Eugen) Köhler, Köhler's *Medizinal-Pflanzen*
in naturgetreuen Abbildungen mit kurz erläuterndem Texte: Volume II.
Edited by G. (Gustav) Pabst (Gera-Untermhaus: Fr. Eugen Köhler, 1887), Plate 157.
Illustrations by Walther Müeller and C. F. Schmidt. Chromolithography by K. Gunther.

p. 200 The white pulp of cacao seeds is colonized by microbes.
Photograph, 2016.

p. 201 Aztec peoples made a frothy chocolate beverage.
Nahua noblewoman preparing chocolate drink. Illustration from Códice (Codex)
Tudela, a codex or cultural encyclopedia of Aztec culture assembled in the 1500s.
Central America, ca. 1530–1554. Museo de América, Madrid.

7. There Is Life at the Edge of Sight

p. 204   A band of our Milky Way galaxy.
Photograph, 2016.

p. 206   The Hooker Telescope at the Mt. Wilson Observatory, California.
Photograph by Edison Hoge, ca. 1940. Image courtesy of the Observatories of the
Carnegie Institution for Science Collection at the Huntington Library, San Marino,
California (COPC 387).

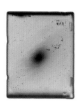

p. 207   Hubble's glass photographic plate.
Plate No. H335H ("Hooker plate 335 by Hubble"), October 6, 1923.
Courtesy of the Carnegie Observatories, Pasadena, California.

pp. 208–209   The Andromeda galaxy.
Image captured by the Spitzer Space Telescope. NASA/Jet Propulsion Laboratory/
California Institute of Technology.

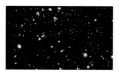

pp. 210–211   The Extreme Deep Field from the Hubble Space Telescope.
Composite image from NASA, the European Space Agency, G. Illingworth,
D. Magee, P. Oesch; the University of California, Santa Cruz; R. Bouwens, Leiden
University; and the HUDF09 Team. NASA/European Space Agency/G. Illingworth
(UCO/Lick Observatory and the University of California, Santa Cruz), R. Bouwens
(UCO/Lick Observatory and Leiden University), and the HUDF09 Team.

p. 212   Inside of a typical galaxy packed with planetary systems.
Artist's impression. NASA/European Space Agency/M. Kornmesser
(European Southern Observatory)

pp. 214–215 Five Earth-like exoplanets.

Artist's conception based on data from the *Kepler* spacecraft, named after Johannes Kepler and specifically tuned to detect an Earth-sized planet orbiting distant stars. Image from NASA/Ames Research Center/NASA Jet Propulsion Laboratory/California Institute of Technology.

p. 216 Archaeal cells from Yellowstone seen in red with ultra-small green *Nanopusillus* ecotosymbionts.

Modified from figure 1(a) of L. Wurch et al. 2016. Genomics-informed isolation and characterization of a symbiotic Nanoarchaeota system from a terrestrial geothermal environment. *Nature Communications* 7: 12115. (CC-BY-4.0)

p. 218 Halophilic archaea can survive for millennia trapped inside halite salt crystals.

Photograph, 2015.

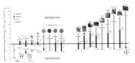

pp. 220–221 The size scales of life.

The images shown are found throughout earlier chapters: several bacteria (*Pelagibacter, Prochlorococcus, Bacillus subtilis, Pseudomonas aeruginosa, Thiomargarita, Streptomyces, Myxococcus xanthus,* cyanobacteria), archaea (haloarchaea), fungi (*Penicillium, Agaricus bisporus,* and *Armillaria ostoyae*); other eukaryotes (*Paramecium, Emiliania huxleyi, Physarum,* and *Ostreococcus tauri*); and animals (*Hypsibius* tardigrades, *Tyrophagus* cheese mites, leaf-cutter ants, and humans). Most are photographed by Scott Chimileski. Satellite imagery is from the NASA Earth Observatory and the National Park Service (Grand Prismatic Spring), and the cyanophage and ectosymbiont images are from M. B. Sullivan et al. 2005. Three *Prochlorococcus* cyanophage genomes: signature features and ecological interpretations. *PLoS Biology* 3(5):e144, and from L. Wurch et al. 2016. Genomics-informed isolation and characterization of a symbiotic Nanoarchaeota system from a terrestrial geothermal environment. *Nature Communications* 7: 12115. All sizes plotted are approximate and based on published size data.

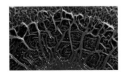

pp. 222–223 Biofilms are macroscopic manifestations of microbes.

Pseudomonas aeruginosa. Composite photograph, 2016.

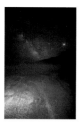

p. 225   Macroscopic manifestations of thermophilic microbes.
Yellowstone National Park, Wyoming. Photograph, 2016.

p. 227   An artistic view inside a *Mycoplasma* cell.
Watercolor painting by David S. Goodsell, the Scripps Research Institute, 2011.

p. 228   Suspected water plumes captured by the Hubble Telescope.
Composite image from NASA/European Space Agency/W. Sparks (STScI)/
United States Geological Survey Astrogeology Science Center.

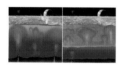

p. 229   Water plumes originate from an ocean created by thermal activity beneath the icy surface.
NASA Jet Propulsion Laboratory. Artwork by Michael Carroll.

pp. 230–231   The ice-covered, cracked surface of Europa.
Composite image from NASA/Jet Propulsion Lab/California Institute of Technology/
SETI Institute.

p. 232   The next-generation James Webb Space Telescope.
Photograph from NASA/Goddard Space Flight Center/Chris Gunn.

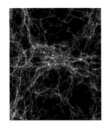

p. 233   The cosmic web: a large-scale structure of the universe that evolves over time.
Image from M. Boylan-Kolchin et al. 2009. Resolving cosmic structure formation with the Millennium-II Simulation. *Monthly Notices of the Royal Astronomical Society* 398(3): 1150–64.

p. 235   Composite image of Earth's western hemisphere.
NASA images by Reto Stöckli, based on data from NASA and NOAA.

HOW TO PHOTOGRAPH MICROBES

p. 238   Green algae from the genus *Pediastrum*.
Ernst (Heinrich Philipp August) Haeckel, *Kunstformen der Natur* (Leipzig und Wien: Verlag des Bibliographischen Instituts, 1904), Plate 34 (labeled *Melethallia*).

p. 239   A macroscopic structure formed by myxobacteria, illustrated by Roland Thaxter, 1892.
Roland Thaxter, On the Myxobacteriaceae, a new order of Schizomycetes. *Botanical Gazette* 17(12):389–406 (1892). Plate XXIV, Figure 24. Photograph of original drawings accessed at the Harvard University Herbaria & Libraries.

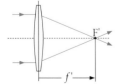

p. 240   Light refraction through a simple glass lens.
The simple lens shown here is a biconvex lens, having two convex faces. This is just one type of simple lens among others like plano-convex (one-sided convex), meniscus, and concave lenses, all of which are used for certain applications within modern optical imaging systems. Diagram by JiPaul, 2014. (CC-BY-SA-3.0).

p. 241 A Yellowstone Mud Volcano.
  Yellowstone National Park, Wyoming. Photograph, 2016.

p. 242 Long-exposure photograph of a Yellowstone Geyser field at night.
  Biscuit Basin, Yellowstone National Park, Wyoming. Photograph, 2016.

p. 243 (above) An alien microbial landscape of barely visible liquid droplets.
  Photograph, 2015.

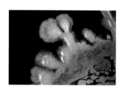

p. 243 (below) Migrating bacterial colonies.
  Photograph, 2014.

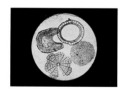

p. 245 One of the first-ever photomicrographs.
  William Henry Fox Talbot, 1840.

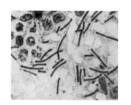

p. 246 One of the first photomicrographs of bacterial cells.
  Bacillus anthracis, photomicrograph. Dr. (Robert) Koch. Method for the Investigation
  for Preserving and Photographing the Bacteria, Plate XVI, Figure 5, in *Contributions
  to Biology of Plants: Volume II, Part III*, edited by Dr. Ferdinand Cohn (Breslau: J. U.
  Kern's Verlag (Max Müller), 1877), pp. 399–434.

p. 247 The Leitz Panphot microscope of the mid-twentieth century.
  Leitz PANPHOT Instruction Book, Figure 1. Ernst Leitz GmbH, Wetzlar, Germany.
  This photograph of the microscope comes from the original manual for the Panphot,
  which may be downloaded in full online.

p. 250 (left)   A beam of blue light excites a biofilm grown within a glass-chambered slide.
Photograph, 2016.

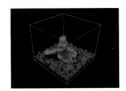

p. 250 (right)   Fluorescence microscopy of van Gogh bundles formed by the bacterium *Bacillus subtilis.*
Micrograph by Jordi van Gestel. For more details, see J. van Gestel, H. Vlamakis, and R. Kolter. 2015. From cell differentiation to cell collectives: *Bacillus subtilis* uses division of labor to migrate. *PloS Biology* 13 (4):e1002141.

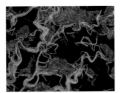

p. 251   A three-dimensional model of an archaeal biofilm produced with confocal microscopy.
Micrograph / 3D model collected at the Center for Biofilm Engineering in Bozeman, Montana.

p. 252   A modern scanning electron microscope and its controls.
Photograph of a Zeiss model instrument kept at the Harvard Center for Nanoscale Systems in 2016.

p. 253 (left)   Macro photograph of a cross section through a squash plant stem.
Plant sample prepared by Lori Shapiro. Photograph, 2015.

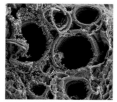

p. 253 (right) Bacteria appear within the squash stem's xylem vascular tissues.
   Plant sample prepared by Jorge Rocha. Scanning electron micrograph, 2016.

p. 254 (left) The species of bacteria inside the plant is *Erwinia tracheiphila*.
   Scanning electron micrograph, 2016.

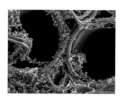

p. 254 (right) Individual *Erwinia* cells and their extracellular matrix seen inside the xylem.
   Scanning electron micrograph, 2016.

p. 255 Photographing the Grand Prismatic Spring.
   Yellowstone National Park, Wyoming. Photograph, 2016.

p. 256 A wild plasmodial slime mold.
   Concord, Massachusetts, near Walden Pond. Photograph, 2016.

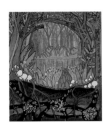

p. 258 Painting by Scott Chimileski.

p. 290 *Hot Springs of the Yellowstone* by Thomas Moran.
Oil on canvas, 1872. Gift of Beverly and Herbert M. Gelfand (M.84.198).
Los Angeles County Museum of Art (LACMA) Image Library.

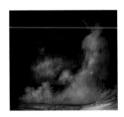

p. 322 Mud Cauldron, Yellowstone National Park, Wyoming.
Photograph, 2016.

p. 362 The slime mold *Physarum polycephalum*.
Photograph, 2016.

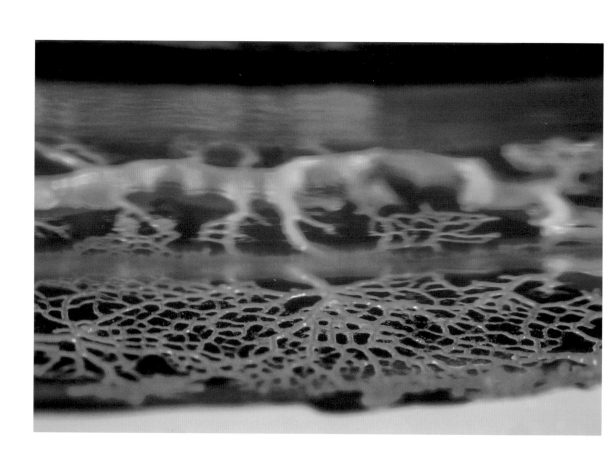

INDEX